# SEMTherm 2017: International Conference on Microtechnology & Thermal Problems in Electronics (MICROTHERM 2017) and International Conference on Smart Engineering of New Materials (SENM 2017)

Journal of Physics: Conference Series Volume 1033

Lodz, Poland
26 - 29 June 2017



**Editors:**

**Ewa Raj**
**Katarzyna Znajdek**

ISBN: 978-1-5108-6408-5
ISSN: 1742-6588

**Printed from e-media with permission by:**

Curran Associates, Inc.
57 Morehouse Lane
Red Hook, NY 12571

**Some format issues inherent in the e-media version may also appear in this print version.**



Printed by Curran Associates, Inc. (2018)

For permission requests, please contact the Institute of Physics at the address below.

Institute of Physics
Dirac House, Temple Back
Bristol BS1 6BE UK

Phone: 44 1 17 929 7481
Fax: 44 1 17 920 0979

techtracking@iop.org

**Additional copies of this publication are available from:**

Curran Associates, Inc.
57 Morehouse Lane
Red Hook, NY 12571 USA
Phone: 845-758-0400
Fax: 845-758-2633
Email: curran@proceedings.com
Web: www.proceedings.com

# Table of contents

## Volume 1033

**SEMTherm 2017: International Conference on Microtechnology & Thermal Problems in Electronics (MICROTHERM2017) and International Conference Smart Engineering of New Materials (SENM2017)**

**26–29 June 2017, Lodz, Poland**

**Accepted papers received: 17 May 2018**
**Published online: 1 June 2018**

### Preface



### Papers

IOP Conf. Series: Journal of Physics: Conf. Series **1033** (2018) 011001 doi:10.1088/1742-6596/1033/1/011001

**PREFACE**

The issue covers selected papers presented during SEMTherm 2017 - the Joint Event of two International Cyclic Conferences: MicroTherm – Microtechnology and Thermal Problems in Electronics and SENM – Smart Engineering of New Materials. The Duoconference was held in Lodz, Poland on 26-29 June 2017. It gathered about 150 participants from 16 countries in Lodz – beautiful, post-industrial city located in the centre of Poland.

MicroTherm is an International Conference on Microtechnology and Thermal Problems in Electronics being organised every two years, since 1996. The success of the first seminar devoted to thermal management aspects, and the successive conferences has led us to the twelve edition. Since the first meeting, the scope of the Conference has expanded, following the electronics progress. Now, it covers the subjects connected with extreme temperature electronics, sensors and measurement techniques, modelling, simulation, wide band-gap materials, packaging and reliability, renewable energy sources and photonics with special emphasis on microelectronic technologies. Traditionally, the programme of MicroTherm 2017, despite of invited talks and regular sessions in a form of planar discussions and poster presentations, included also Students' Session. The latter gave an opportunity for students and young researchers to present their first achievements in the field of science.

The tradition of SENM dates back to 1985 when the First International Conference on Diamond Crystallization under Reduced Pressure was held in Jabłonna in Poland. Since then, the Conference has been organised regularly every five years, and from 2015, it is the biennial event. From the early beginning, the main aim of this conference has been the creation of the international forum of scientists for the presentation of the newest achievements and investigation results in vacuum and plasma techniques, synthesis, characterisation and application of diamond, its derivatives and other wide bandgap materials. Over time, the scope of the conference has expanded to include nanomaterials, biomaterials and new functional materials, in the range of production, design and optimisation of the synthesis and modification technologies.

Owing to the overlapping scopes of MicroTherm and SENM conferences, organisation of the Joint Event SEMTherm has been widely appreciated. Details about the forthcoming event will be available at the duoconference website soon.

Ewa Raj and Katarzyna Znajdek
Editors

IOP Conf. Series: Journal of Physics: Conf. Series **1033** (2018) 011002 doi:10.1088/1742-6596/1033/1/011002

# Peer review statement

All papers published in this volume of *Journal of Physics: Conference Series* have been peer reviewed through processes administered by the proceedings Editors. Reviews were conducted by expert referees to the professional and scientific standards expected of a proceedings journal published by IOP Publishing.

IOP Conf. Series: Journal of Physics: Conf. Series **1033** (2018) 012001 doi:10.1088/1742-6596/1033/1/012001

# Thermal model of the IGBT module

**Paweł Górecki, Krzysztof Górecki, Janusz Zarębski**

Gdynia Maritime University, Department of Marine Electronics, Morska 83, Gdynia, Poland

e-mail: k.gorecki@we.am.gdynia.pl

**Abstract**. In the paper the dynamic compact thermal model of the IGBT module is proposed. This model makes it possible to calculate internal temperatures of diodes and transistors included in the considered IGBT module with selfheating in each device and mutual thermal couplings between them taken into account. The form of this model and the method of estimation of its selected parameters are described. Some results of measurements and calculations illustrating the usefulness of this model are shown and discussed.

## 1. Introduction

The aspiration to miniaturize electronic and power electronic equipment causes that more and more subcircuits of these equipment are realized as integrated circuits. Power modules are commonly used in switch-mode power supplies [1, 2]. These modules contain in the common case semiconductor devices indispensable to construct one branch of inverters [2]. Typically there are two transistors and two diodes. From the point of view of technology such a module can be treated as a hybrid integrated circuit [3].

During the operation of semiconductor devices contained in power modules, heat is generated [4, 5]. The considered devices are situated on the common basis near each other, which results in thermal interactions between them [6], and finally non-uniform distribution of temperature in such a module is observed [7]. For the purpose of calculating temperature distribution in the considered module estimation of the internal temperature of every semiconductor device is desirable [3, 4, 5]. To realize this task it is necessary to use the thermal model of the power module. Such a model should take into account both self-heating in each semiconductor device and mutual thermal coupling between these devices.

The problem of modelling thermal properties of electronic devices is considered in many papers [4, 8, 9, 10, 11, 12, 13, 14]. The cited papers are devoted to thermal properties of discrete devices. In turn, the problem of modelling mutual thermal couplings between semiconductor devices are described e.g. in [3, 5, 15].

In the paper the compact thermal model of the IGBT module Power Sem PSI 25/06 and the manner of estimating the parameters values of this model are proposed. The correctness of the model is verified experimentally.

## 2. Investigated module

The power module used in investigations contains two IGBTs and two diodes connected electrically into the branch of the inverter [16]. The diagram of this module is shown in Fig. 1a.

IOP Conf. Series: Journal of Physics: Conf. Series **1033** (2018) 012001 doi:10.1088/1742-6596/1033/1/012001

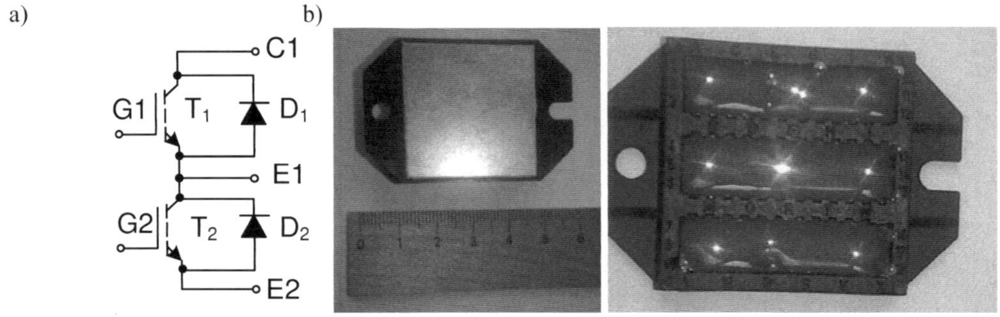


**Figure 1.** Diagram of the IGBT module PSI 25/06 (a) and a view of the case of this module (b)

The considered module has 5 terminals, and transistors contracted in it are characterized by the maximum collector current equal to 18.5 A and the maximum reverse voltage reaching 600 V. With the same maximum conduction current the diodes are characterized, and the catalogue value of their forward voltage amounts to 2.58 V at the current equal to 15 A. The described module is situated in the plastic case of the dimensions 34.3x51x8mm shown in Fig.1b, whereas the top part of the case is the metal heat-sink. The values of the junction-to-case thermal resistance given by the manufacturer are equal to 1.52 K/W for the transistor and 3 K/W for the diode, respectively.

## 3. Model form

As it is visible in Fig. 1, the considered module contains 4 semiconductor devices situated in the common case, but there are four separated semiconductor chips. Therefore the mutual thermal couplings between these devices occur.

In order to calculate the internal temperature of electronic devices their thermal model is indispensable. The thermal model of the considered IGBT module elaborated by the authors belongs to the group of compact thermal models [3, 4, 5, 17]. This model makes it possible to calculate internal temperatures of both the transistors and both the diodes included in the considered IGBT module with a selfheating phenomenon and mutual thermal couplings between all the components of this module taken into account.

The value of the internal temperature of these devices depends on the value of the ambient temperature $T_a$ and the temperature excess caused by self-heating phenomena and by mutual thermal couplings between the semiconductor devices contained in the module. Using the idea of compact thermal modelling [3, 4, 5], the compact thermal model of the IGBT module can be described with convolution integrals [18]. For example, the internal temperature of the transistor $T_1$ can be described with the following equation:

$$T_{jT1}(t) = T_a + \int_0^t Z'_{thT1}(\tau - t) \cdot p_{T1}(\tau) d\tau + \int_0^t Z'_{thT2T1}(\tau - t) \cdot p_{T2}(\tau) d\tau + \int_0^t Z'_{thD1T1}(\tau - t) \cdot p_{D1}(\tau) d\tau + \int_0^t Z'_{thD2T1}(\tau - t) \cdot p_{D2}(\tau) d\tau \quad (1)$$

where $Z'_{thT1}(t)$ denotes the time derivative of self transient thermal impedance of the transistor $T_1$, $Z'_{thT2T1}(t)$, $Z'_{thD1T1}(t)$, $Z'_{thD2T1}(t)$ - time derivatives of mutual transient thermal impedances between the transistors $T_2$ and $T_1$, between the diode $D_1$ and the transistor $T_1$, between the diode $D_2$ and the transistor $T_1$, respectively; whereas $p_{T1}(t)$, $p_{T2}(t)$, $p_{D1}(t)$, $p_{D2}(t)$ - power dissipated in transistors $T_1$ and $T_2$ and in the diodes $D_1$ and $D_2$, respectively.

Internal temperatures of the remaining components of the considered module are described by equations of the form analogous to the equation (1). Therefore, in the description of the thermal model of the considered module four self transient thermal impedances and twelve mutual transient thermal impedances appear. Because the two-way heat transfer between each pair of semiconductor devices exists, the considered module is characterized by the same mutual transient thermal impedances. Hence, only six (instead of 12) mutual transient thermal impedances are needed in the model.

## 4. Estimation of the model parameters values

In order to apply the presented model in practice, it is indispensable to estimate the values of all its parameters. This process is performed in two steps. First, the waveforms of self and mutual transient thermal impedances occurring in the considered model are measured. Next, for each of them the parameters values of the classical $Z_{th}(t)$ model are calculated. This model is given by following equation [3, 4, 5, 18]:

$$Z_{th}(t) = R_{th}\left[1 - \sum_{i=1}^{n} a_i \cdot \exp\left(-\frac{t}{\tau_{thi}}\right)\right] \tag{2}$$

In Eq. (2) $R_{th}$ denotes thermal resistance, $a_i$ - the weight coefficient corresponding to i-th thermal time constant $\tau_{thi}$, while N is the number of thermal time constants. For each self and mutual transient thermal impedance the values of parameters $R_{th}$, $a_i$, $\tau_{thi}$ must be estimated.

Due to electric connections between semiconductor devices contracted in the considered module, special methods of measurements of self and mutual transient thermal impedances have to be used. The measurement method of each transient thermal impedance requires a different measuring set-up, whereas the measurement procedure is identical. The measurement is realized with the use of the indirect electrical method and the conception of the so called cooling curve [19]. Thermally sensitive parameters are the voltage between the gate and the emitter of the transistor at the small constant value of the collector current or the forward voltage of the diode operating with the constant current of the small value. For example, the measurement set-up to measure self transient thermal impedance of the transistor T1 $Z_{thT1}(t)$ is shown in Fig. 2, whereas the set-up to measure mutual transient thermal impedance between the diodes $Z_{thD1D2}(t)$ is presented in Fig. 3.

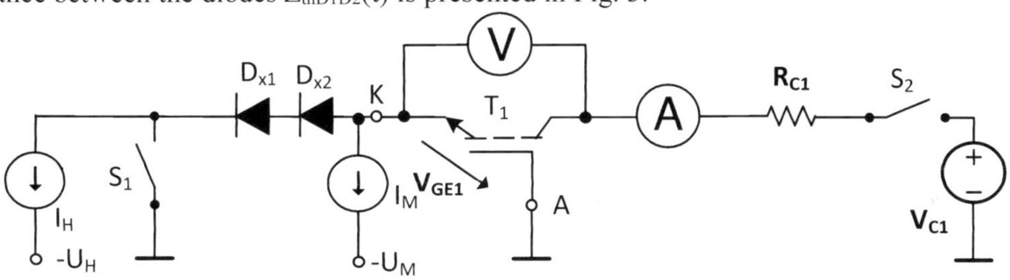


**Figure 2.** Measurement set-up to measure transient thermal impedance $Z_{thT1}(t)$

The measurement of transient thermal impedance of the transistor $T_1$ is realized in four steps. First, thermometric characteristics $V_{GE1}(T)$ of the transistor $T_1$ are measured at the collector current $I_M$. During calibration the tested module (DUT) is situated in the thermostat, the switch $S_1$ is closed and the switch $S_2$ is opened. As a result of calibration the slope of the thermometric characteristic $\alpha_V$ is calculated. The characteristic $V_{GE1}(T)$ is linear only for the selected values of $I_M$ current. The value of this current must be much higher than the sub-threshold current and simultaneously so small that selfheating phenomenon could be omitted.

In the second step the transistor $T_1$ operates in the active normal mode. The switch $S_1$ is opened and the switch $S_2$ is closed. In the transistor the dissipated power is equal to $I_H \cdot V_{CE}$. As a result of selfheating, the internal temperature of this transistor increases until the steady state is obtained. Then, the values of the current $I_H$ and the voltage $V_{CE}$ are measured. At time t = 0 the switch $S_1$ is closing and the switch $S_2$ is opening. Next, the waveform of the voltage $V_{GE1}(t)$ is registered while cooling the tested module until the steady state is obtained. Finally, the waveform of transient thermal impedance is calculated with the formula:

$$Z_{thT1T2}(t) = \frac{V_{GE1}(t) - V_{GE1}(t=0)}{\alpha_V \cdot I_{C1} \cdot V_{CE1}} \tag{3}$$

The measurement of mutual transient thermal impedance between the diode $D_1$ and the diode $D_2$ is realized in four steps. First, thermometric characteristics $V_{D1}(T)$ of the diode $D_1$ are measured at the forward current $I_M$. During calibration the tested module (DUT) is situated in the thermostat, the switch $S_1$ is closed and the switch $S_2$ is opened. As a result of calibration the slope of the thermometric characteristic $\alpha_V$ is calculated.

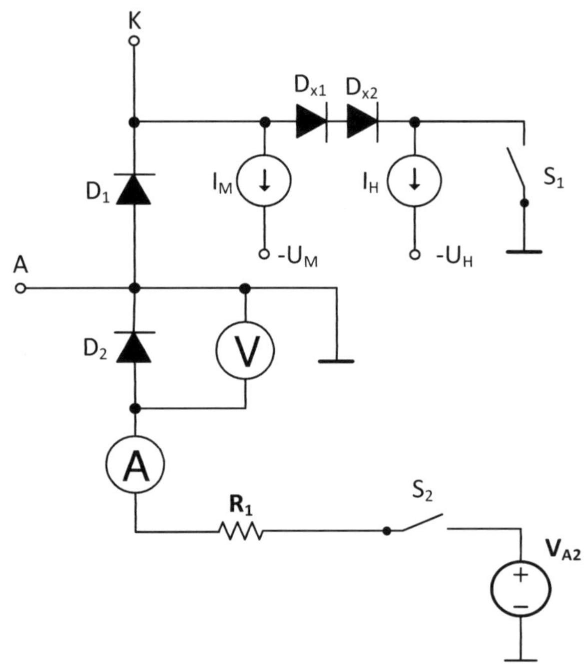


**Figure 3.** Measurement set-up to measure self and mutual transient thermal impedances of the diodes

In the second step the diode $D_2$ operates in the forward mode. The switch $S_1$ is opened and the switch $S_2$ is closed. As a result of mutual thermal couplings the internal temperature of the diode $D_1$ increases due to a selfheating phenomenon occurring in the diode $D_2$ until the steady state is obtained. Then, the values of the current $I_{D2}$ and the voltage $V_{D2}$ are measured. At time t = 0 the switch $S_1$ is closing and the switch $S_2$ is opening. Next, the waveform of the voltage $V_{D1}(t)$ is registered while cooling the tested module until the steady state is obtained. Finally, the waveform of transient thermal impedance is calculated with the formula:

$$Z_{thD1D2}(t) = \frac{V_{D1}(t) - V_{D1}(t=0)}{\alpha_V \cdot I_{D2} \cdot V_{D2}} \tag{4}$$

For all the other thermal parameters the measurement is realized in three steps. The first step includes calibration of thermometric characteristic. The second step - dissipation of the power $P_H$ in one of the components of the module, which plays the role of a heat source until the thermally steady state is obtained. The third step - cooling this element and recording waveforms of the thermally sensitive parameter from the moment t = 0 (switching off the power dissipated in the heat source) until the thermally steady state is obtained. On the basis of the measured waveforms of the thermally sensitive parameter and the power $P_H$, waveforms of transient thermal impedances are calculated using equations of the form similar to Eq. (3) for self transient thermal impedances or similar to Eq. (4) for mutual transient thermal impedances.

Using measured waveforms of self and mutual transient thermal impedances for each of them the values of parameters $R_{th}$, $a_i$ and $\tau_{thi}$ are estimated with the use of the algorithm described in [20]. In this algorithm the value of $R_{th}$ is equal to mean value of the transient thermal impedance at the steady state. In turn, values of parameters $a_i$ and $\tau_{thi}$ are calculated using least square method in approximation transformed Eq. (2) by a linear function starting from the longest thermal time constant.

## 5. Results of measurements and calculations

Using the methods described in Section 4, self and mutual transient thermal impedances existing in the thermal model of the IGBT module were measured. For example, in Fig. 4a the measured (solid lines) waveforms of self transient thermal impedance $Z_{thT1}(t)$ of the transistor $T_1$ and mutual transient thermal impedance $Z_{thT1D1}(t)$ between the transistor $T_1$ and the diode $D_1$ are shown, whereas in Fig. 4b the measured waveforms of self transient thermal impedance $Z_{thD1}(t)$ of the diode $D_1$ and mutual transient impedance $Z_{thD1D2}(t)$ between the diodes are presented. In the same figures the waveform of transient thermal impedances obtained with the use of Eq. (2) are marked with dashed lines.

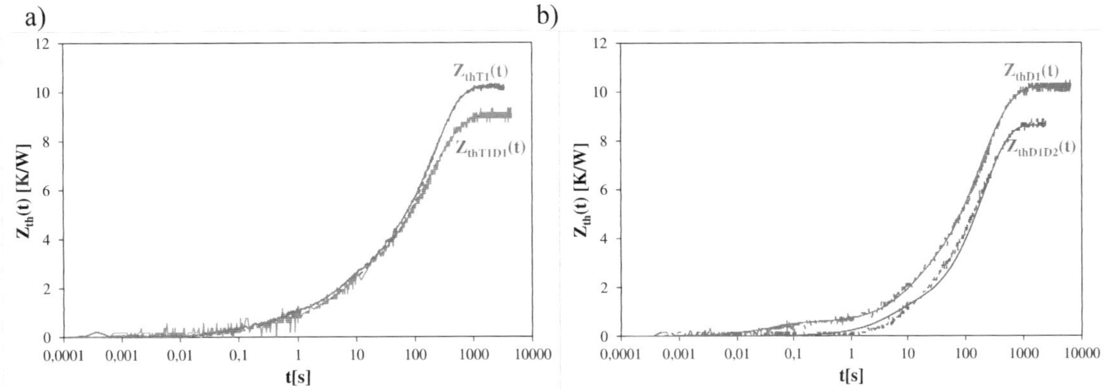


**Figure 4.** Measured and calculated waveforms of the selected self and mutual transient thermal impedances

As it is visible, good agreement between the measured and modelled waveforms was obtained. The differences between results of calculations and measurements do not exceed 0.2 K/W. It is worth noticing that visible differences between the waveforms of self and mutual transient thermal impedances are observed. Particularly, for the waveforms of self and mutual transient thermal impedances $Z_{thT1}(t)$ and $Z_{thT1D1}(t)$ one can observe that differences occur only at the steady state. In turn, the waveform of $Z_{thD1D2}(t)$ is visibly late in comparison with the waveform of $Z_{thD1}(t)$.

In Table 1 the values of parameters describing the considered self and mutual transient thermal impedances by Eq. (2) are collected.

**Table 1.** Parameters values describing the considered self and mutual transient thermal impedances of the considered IGBT module.

| parameter | $Z_{thT1}(t)$ | $Z_{thT1D1}(t)$ | $Z_{thD1}(t)$ | $Z_{thD1D2}(t)$ |
|---|---|---|---|---|
| $R_{th}$ [K/W] | 10.15 | 9.05 | 10.24 | 8.63 |
| $a_1$ | 0.736 | 0.479 | 0.384 | 0.886 |
| $a_2$ | 0.18 | 0.257 | 0.416 | 0.114 |
| $a_3$ | 0.068 | 0.158 | 0.027 | - |
| $a_4$ | 0.014 | 0.053 | 0.12 | - |
| $a_5$ | 0.002 | 0.038 | 0.053 | - |
| $a_6$ | - | 0.015 | - | - |
| $\tau_{th1}$ [s] | 212.3 | 281.9 | 331.6 | 204.6 |
| $\tau_{th2}$ [s] | 6.64 | 78.25 | 123.6 | 4.57 |
| $\tau_{th3}$ [s] | 0.418 | 7.75 | 12.68 | - |
| $\tau_{th4}$ [s] | 0.336 | 2.06 | 9.68 | - |
| $\tau_{th5}$ [ms] | 0.04 | 163.8 | 0.04 | - |
| $\tau_{th6}$ [ms] | - | 54.33 | - | - |

As it is visible, differences between thermal resistance $R_{th}$ of the transistor T1 and the diode D1 are very small. They do not exceed 1%. In turn, mutual thermal resistances are even 15% smaller than self thermal resistances. The steady state of all the considered waveforms is obtained after nearly 1000 s.

The practical usefulness of the elaborated model is illustrated in Fig.5. In this figure the calculated waveforms of internal temperatures of the selected components of the considered module are shown.

The waveforms shown in Fig. 5a were obtained during power dissipation in the transistor T1 only. This power has the shape of a single rectangular impulse of the amplitude equal to 10 W and duration time equal to 1000 s. In turn, in Fig. 5b the waveforms of the internal temperature of the same devices obtained at dissipation of power of the same shape of a rectangular impulse of duration time equal to 1000s in the transistor $T_1$ (amplitude equal to 10 W) and in the diode $D_1$ (amplitude equal to 5 W) simultaneously.

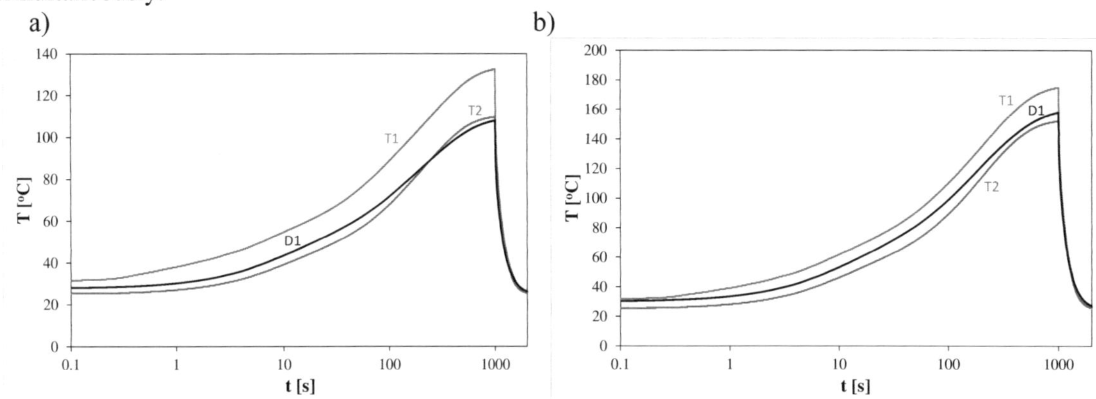


**Figure 5.** Calculated waveforms of the internal temperatures of components of the considered module at different waveforms of the power dissipated in these components

As it is visible, the internal temperature of all the considered devices changes while heating and cooling the used heat sources. This is a result of mutual thermal couplings between the components of the IGBT module. In both the considered cases the highest value of temperature is observed for the transistor $T_1$, in which the power of the highest amplitude is dissipated. The observed differences in the maximum value of the components internal temperature reach even 30°C and this value is smaller, when the power is dissipated in two devices simultaneously.

## 6. Conclusions

In the paper the compact thermal model of the IGBT module containing IGBTs and diodes was proposed. This model makes it possible to calculate waveforms of the internal temperature of each semiconductor device contracted in this module taking into account both self-heating phenomena in each of these devices and mutual thermal couplings between all the devices. The analytical form of this model and the procedure of estimation of the parameters values of this model are presented.

The measured waveforms of self and mutual transient thermal impedances differ from each other even by 15% and the classical analytical formula makes it possible to accurate approximate results of measurements. It is proved that both selfheating in each component of the module and mutual thermal coupling between them must be taken into account in order to model correctly thermal phenomena in the considered module.

The presented results of calculations illustrate that differences in the values of internal temperatures of the components of the considered module can reach even 30°C. This proves that strong non-uniform temperature distribution occurs in the considered module.

In further investigations some infrared measurements of the considered module operating without this case will be performed in order to measure temperature distribution in this module.

**Acknowledgements**

The scientific work financed with the Polish science budget resources in the years 2017-2021, as the investigation project within the framework of the program "Diamentowy Grant".

**References**

[1] Rashid MH 2007 Power Electronic Handbook *Acad. Press, Elsevier* (New York).

[2] Perret R 2009 Power Electronics Semiconductor Devices *John Wiley & Sons, Ltd.* (London).

[3] Górecki K, Ptak P 2015 Modelling mutual thermal coupling in LED modules *Microelectronics International* **32** (3) 152-157.

[4] Szekely V 1997 A New Evaluation Method of Thermal Transient Measurement Results *Microelectronics Journal* **28** (3) 277-292.

[5] Górecki K 2015 Modelling mutual thermal interactions between power LEDs in SPICE *Microelectronics Reliability* **55** (2) 389-395.

[6] Yener Y, Kakac S 2008 Heat Conduction *Taylor &Francis* (Washington).

[7] Górecki P 2017 Investigation of the influence of thermal phenomena on characteristics of IGBTs contained in power modules *24$^{th}$ International Conference Mixed Design of Integrated Circuits and Systems Mixdes 2017*, Bydgoszcz 355-359.

[8] Bouguezzi S, Ayadi M, Ghariani M 2016 Developing a simplified analitical thermal model of multi-chip power module *Microelectronics Reliability* **66** 64-77.

[9] Sofia JW 1995 Analysis of thermal transient data with synthesized dynamic-models for semiconductor-devices *IEEE Transactions on Components Packaging and Manufacturing Technology Part A* **18** (1) 39-47.

[10] Bagnoli PE, Casarosa C, Ciampi M, Dallago E 1998 Thermal resistance analysis by induced transient (TRAIT) method for power electronic devices thermal characterization *IEEE Transactions on Power Electronics, I. Fundamentals and Theory* **13** (6) 1208-19.

[11] Górecki K, Zarębski J 2010 Nonlinear compact thermal model of power semiconductor devices *IEEE Transactions on Components and Packaging Technologies* **33** (3) 643-647.

[12] Bernardoni M, Delmonte N, Cova P, Menozzi R 2010 Thermal modeling of planar transformer for switching power converters *Microelectronics Reliability* **50** (9-11) 1778-1782.

[13] Janke W 1992 Zjawiska termiczne w elementach i układach półprzewodnikowych *WNT* (Warszawa).

[14] Luo ZH, Ahn H, El Nokai MA 2004 A thermal model for insulated gate bipolar transistor module *IEEE Transactions on Power Electronics* **19** (4) 902-907.

[15] Ye J, Yang K, Ye HZ, Emadi A 2017 A fast electro-thermal model of traction inverters for electrified vehicles *IEEE Transactions on Power Electronics* **32** (5) 3920-3934.

[16] IGBT Module PSI 25/06 2002 Catalogue data *Powersem* (Schwabach).

[17] Górecki K, Górecki P 2017 Modelling dynamic characteristics of the IGBT with thermal phenomena taken into account *Microelectronics International* **34** (3) 160-164

[18] Zarębski J, Górecki K 2007 Properties of Some Convolution Algorithms for the Thermal Analysis of Semiconductor Devices *Applied Mathematical Modelling* **31** (8) 1489 – 1496.

[19] Blackburn DL 2004 Temperature Measurements of Semiconductor Devices – A Review *20$^{th}$ IEEE Semiconductor Thermal Measurement and Management Symp. SEMI-THERM* San Jose 70-80.

[20] Górecki K, Rogalska M, Zarębski J 2014 Parameter estimation of the electrothermal model of the ferromagnetic core *Microelectronics Reliability* **54** (5) 978-984.

IOP Conf. Series: Journal of Physics: Conf. Series **1033** (2018) 012002 doi:10.1088/1742-6596/1033/1/012002

# Numerical analysis and optimization of $Cu_2O/TiO_2$, $CuO/TiO_2$, heterojunction solar cells using SCAPS

**Paulina Sawicka-Chudy[1] , Maciej Sibiński[2], Grzegorz Wisz[3], Elżbieta Rybak-Wilusz[4], Marian Cholewa[1]**

[1]Department of Biophysics, Faculty of Mathematics and Natural Sciences, University of Rzeszow, 1 Pigonia Street, 35-317 Rzeszow, Poland.
[2]Department of Semiconductor and Optoelectronic Devices, Lodz University of Technology, Wólczańska 211/215, 90-924 Lodz, Poland.
[3]Department of Experimental Physics, Faculty of Mathematics and Natural Sciences, University of Rzeszow, Pigonia 1 Street, Rzeszow 35-317, Poland.
[4]Department of Environmental Engineering, Faculty of Civil and Environmental Engineering and Architecture, Rzeszow University of Technology, 12 Powstańców Warszawy Street, 35-959 Rzeszow, Poland.

E-mail: sawicka61@wp.pl

**Abstract**. In the presented work, the $Cu_2O/TiO_2$ and $CuO/TiO_2$ heterojunction solar cells have been analyzed by the help of Solar Cell Capacitance Simulator (SCAPS). The effects of various layer parameters like thickness and defect density on the cell performance have been studied in details. Numerical analysis showed how the absorber (CuO, $Cu_2O$) and buffer ($TiO_2$) layers thickness influence the short-circuit current density (Jsc) and efficiency ($\eta$) of solar cells. Optimized solar cell structures of $Cu_2O/TiO_2$ and $CuO/TiO_2$ showed a potential efficiency of ~9 and ~23%, respectively, under the AM1.5G spectrum. Additionally, external quantum efficiency (EQE) curves of the $CuO/TiO_2$ and $Cu_2O/TiO_2$ solar cells for various layers thickness of $TiO_2$ were calculated and the optical band gap ($E_g$) for CuO and $Cu_2O$ was obtained. Finally, we examined the effects of defect density on the photovoltaic parameters.

## 1. Introduction

Semiconducting metal oxides have been a focus of fundamental research recently [1–4]. $Cu_2O/TiO_2$, $CuO/TiO_2$ p–n junction have emerged as promising materials for optoelectronics and these materials are the possible solution for cheap and competitive solar cells construction. In the literature there are many reports on solar cells based on copper oxide as active layers. The conversion efficiency is the most important property in the PV domain [5,6]. Although the theoretical limit of the energy conversion efficiency of a $Cu_2O$ solar cell is ~ 20% [7], the highest efficiency obtained on substrates of $Cu_2O$ is 3.83% [8]. Table 1 shows values of solar cells efficiencies published in the literature. $TiO_2$ layers were fabricated by anodic oxidation of a Ti foil [9], sputtering [10], hydrothermal synthesis [8], and sintering of a $TiO_2$ nanoparticle paste [11], whereas $Cu_2O$ layers were deposited by electrodeposition and sputtering [12].

In this work, we present a numerical study of the thin film solar cells $Cu_2O/TiO_2$ and $CuO/TiO_2$ with Solar Cell Capacitance Program. It is used to calculate the photovoltaic parameters, with the most important: short-circuit current density (Jsc) and efficiency ($\eta$) under standard illumination (AM 1.5G, 100 mW/cm$^2$, 300K).

The purpose is to study the influence the construction of the active layers on the performance of the solar cells. Towards this goal current density (Jsc) and efficiency ($\eta$) characteristics have been calculated in different thickness layer materials. Additionally, we examined the quantum efficiency

(EQE) curves in various thickness of $TiO_2$ and the effects of defect density on the photovoltaic parameters.

**Table 1.** Values of efficiency and heterojunction solar cells on the basis of [13-16].

| Heterojunction solar cell | Efficiency [%] |
|---|---|
| $Cu_2O/ZnO$ | 0.40% [13] |
| ZnO (ZO)/$Cu_2O$ | 3.83% [8] |
| $TiO_2/Cu_2O$ | 1.25%[14] |
| $TiO_2/Cu_2O$ | ~0.01% [15] |
| $TiO_2/CuO$ | 0.14%, [16] |
| $TiO_2/CuO/Cu$ | 1.62% [16] |

## 2. The simulation program SCAPS

There exists several numerical simulation tools for thin film solar cells [17] e.g. AFORS-HET [18], AMPS [19], ASPIN [20], SCAPS [21]. In the present investigation, we have used SCAPS-1D simulation software to study the $Cu_2O/TiO_2$, $CuO/TiO_2$ solar cells. There are some examples of successful thin film solar cell optimization (Frisk et al. [22]) and Zhao et al. [23]) like by layer thickness, carrier density, defect density etc. adjustment, using AMPS-1D simulator.

SCAPS is a Windows application program, developed at the University of Gent with LabWindows/CVI of National Instruments. It has been made available to university researchers in the photovoltaic community after the second PV World Conference in Wien, 1998 [24]. SCAPS version 3.2.00 is available now. Computer simulation is a valuable tool that can provide useful information for the screening process of materials to be tested experimentally [25].

The numerical simulation program SCAPS has been designed to simulate CIGS and CdTe-based thin film solar cell devices [26]. The user can calculate results in the form of following characteristics: I-V, C-V, C-f, Q(λ), band diagrams, electric field, carrier densities, partial recombination currents.
The user can set parameters of materials and an operating point: temperature, voltage, frequency and illumination [27]. Device is represented as a stack of layers, up to 7 semiconductor layers with specified properties, separate entries for interface parameters and two additional layers for front and back contacts [28]. All layers are characterized by numerical description of parameters.

## 3. $Cu_2O$, CuO and $TiO_2$ properties

Copper forms two well-known stable oxides: cupric oxide (CuO) and cuprous oxide ($Cu_2O$). These two materials have different physical properties, different colors, crystal structures and electrical properties [29, 30]. Both forms are a natural p-type semiconductors [31].

The first attempt to use $Cu_2O$ for PV application was done by National Science Foundation and Joint Centre in 1978 [25, 32]. Cuprous oxide has low electron affinity (3.2 eV) [25, 33, 34], and very high hole mobility [25,35]. Thus, it is suggested as a potential hole transport material in heterojunction based solar cells [25, 36]. A few attempts were made to fabricate n-type $Cu_2O$ and homo-junction $Cu_2O$ cells, but the achieved efficiency is very low so far [37].

Titanium dioxide ($TiO_2$) belongs to the transition metal oxides, and is one of the best studied materials [38] in the last decade owing to its remarkable optical and electronic properties. Titanium dioxide presents good durability and a high refractive index; therefore it is suitable for applications such as antireflection coating, multilayer optical coating and optical wave-guides [39].
Properties of the materials, used in the simulations are available in the literature. The parameters of $Cu_2O$, CuO and $TiO_2$ used in our simulations are listed down in Table 2.

IOP Conf. Series: Journal of Physics: Conf. Series **1033** (2018) 012002 doi:10.1088/1742-6596/1033/1/012002

**Table 2.** The parameters values of $Cu_2O$, CuO and $TiO_2$ at 300 K on the basis of [25, 40-54].

| Material properties | n-$TiO_2$ | p-$Cu_2O$ | p-CuO |
|---|---|---|---|
| **Band gap [eV]** | 2.26 | 2.17 | 1.51 |
| **Electron affinity [eV]** | 4.20 | 3.20 | 4.07 |
| **Dielectric permittivity (relative)** | 10.00 | 7.11 | 18.10 |
| **CB (conduction band) effective density of states [1/cm$^3$]** | 2.0E+17 | 2.0E+17 | 2.2E+19 |
| **VB (valence band) effective density of states [1/cm$^3$]** | 6.0E+17 | 1.1E+19 | 5.5E+20 |
| **Electron mobility [cm$^2$/Vs]** | 1.0E+2 | 2.0E+2 | 10.0E+1 |
| **Hole mobility [cm$^2$/Vs]** | 25.0 | 8.0E+1 | 1.0E-1 |
| **Shallow uniform donor density [1/cm$^3$]** | 1.0E+17 | 0 | 0 |
| **shallow uniform acceptor density [1/cm$^3$]** | 0 | 1.0E+18 | 1.0E+16 |
| **Advantages** | good transmission in the visible and near infrared regions, good adhesion, high stability against mechanical abrasion, chemical attack, high temperatures | high absorption coefficient, material abundance, non-toxicity, and low-cost fabrication | abundant, nonhazardous source materials, it can be prepared by low cost solution methods |

## 4. The thickness effects on the photovoltaic parameters

Absorber layer is the most important component of the solar cell where the incident photons are absorbed and excess carriers are generated. The primary role of n-buffer layer is to form a p-n junction with the p-absorber layer. The thickness of buffer layer should be as small as possible to reduce the series resistance of the PV device [55].

In the present study, the effect of absorber and buffer layers thickness on the cell efficiency (η) and short circuit current ($J_{sc}$) are simulated by changing the thickness from 1.0 to 10.0 μm ($Cu_2O$, CuO) and from 0.1 to 0.6 μm ($TiO_2$) without introducing additional defects (Figs. 2a,b, 3a,b). The proposed layer thickness is provided by the technological conditions of the thin film manufacturing technologies. A schematic structure of $Cu_2O/TiO_2$ and $CuO/TiO_2$ is shown in Fig. 1.

**Figure 1.** Schematic of heterojunction $TiO_2/CuO(Cu_2O)$ solar cell structure with SCAPS program.

a) b)

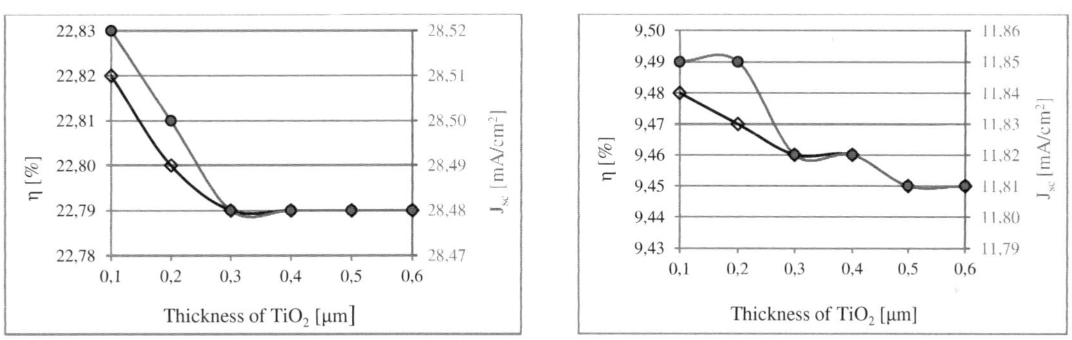


**Figure 2.** Conversion efficiency and short circuit current as function of buffer layer thickness for: $CuO/TiO_2$ a) and $Cu_2O/TiO_2$ b).

To the $CuO/TiO_2$ structure, the cell efficiency drops with an increase in $TiO_2$ layer thickness up to 0.3 μm. With further increase in buffer layer thickness, the efficiency remains nearly constant. This corresponds to a cell efficiency of 22.78%. For $Cu_2O/TiO_2$ the cell efficiency varies differently in relation to $TiO_2$ layer thickness, it decreases slightly and then remains constant and again drops with an increase in buffer layer thickness. Hence, for the present study we have chosen the optimal thickness of $TiO_2$ buffer layer to be 0.3 μm for $Cu_2O/TiO_2$ and for $CuO/TiO_2$. This corresponds to a cell efficiency of 22.78% and 9.46% respectively.

It is important to note that the curves have a similar trend in Fig.3 a,b. The solar cell efficiency increases initially with increase in CuO, $Cu_2O$ layers thickness and nearly saturates at higher values. The results are in agreement with the earlier reported work [55]. When the thickness of the CuO and $Cu_2O$ layer increases, the layers can absorb more photons, resulting in increased photogenerated current. Which are mainly attributed to the enhanced absorption of incident light due to the thickness of the absorber layer [56]. This saturation in efficiency is because of increased probability of SRH (Shockley-Read-Hall) recombination (due to finite carrier diffusion length) with increase in absorber thickness. Moreover, thicker absorber layer means higher material cost and fabrication cost [55]. Therefore, for the presented investigation, the optimized thickness for CuO, $Cu_2O$ absorber was stated at the level of 4.0 μm. The corresponding cells efficiencies are 9.48% and 22,65% respectively.

IOP Conf. Series: Journal of Physics: Conf. Series **1033** (2018) 012002 doi:10.1088/1742-6596/1033/1/012002

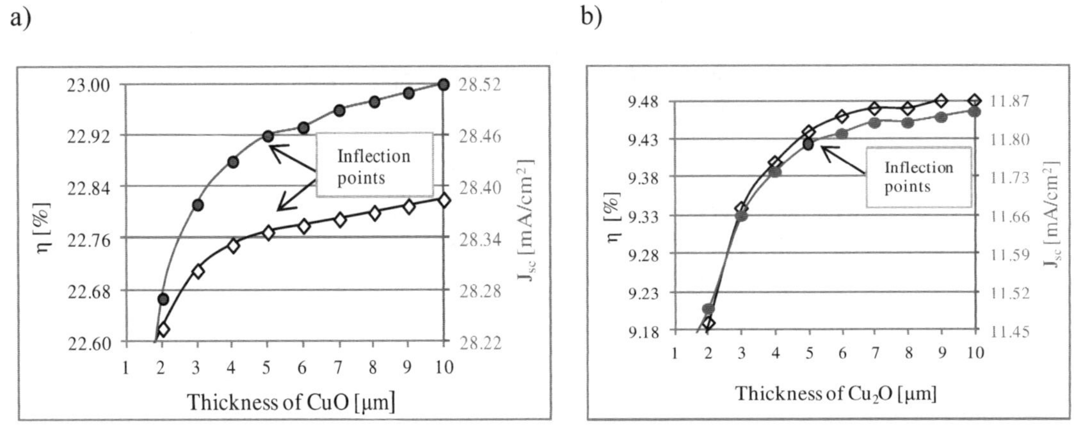


**Figure 3.** Conversion efficiency and short circuit current as function of buffer layer thickness for: $CuO/TiO_2$ a) and $Cu_2O/TiO_2$ b).

The variation of absorption coefficient (α) for CuO and $Cu_2O$ layers was depicted in Fig. 4. The layers absorbs most of the visible light below 500 nm for $Cu_2O$ and 600 nm for CuO. In the wavelength range of 500–570 for $Cu_2O$ and 600–800 nm for CuO the absorption spectra decreases slightly and approaches the band gap near 600 and 820 nm respectively.

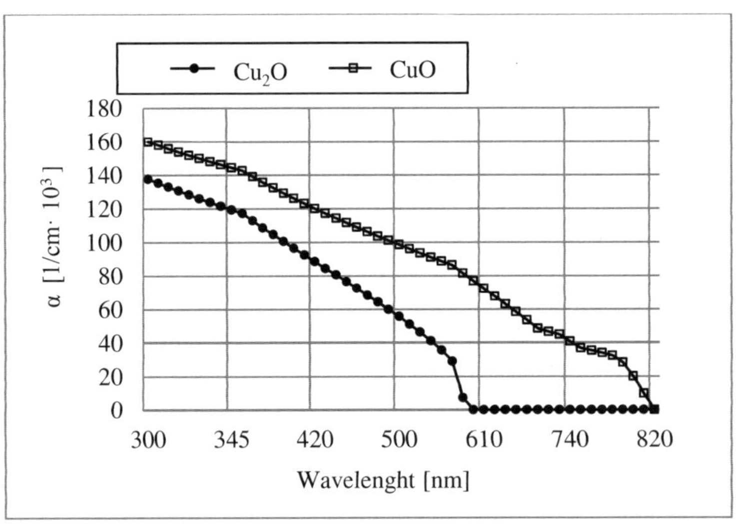


**Figure 4.** Absorption coefficient for CuO and $Cu_2O$ thin films.

## 5. Quantum efficiency QE analysis

Fig. 5 presents the external quantum efficiency (EQE) curves of the $CuO/TiO_2$ (a) and $Cu_2O/TiO_2$ (b) solar cells for various layers thickness of $TiO_2$ (x) in the range 0.1÷0.6 μm. The defect density was stated for $5 \cdot 10^{15}/cm^3$ for CuO and $Cu_2O$, and $1 \cdot 10^{20}/cm^3$ for $TiO_2$. The obtained simulations reveal that the $CuO/TiO_2$ and $Cu_2O/TiO_2$ exhibit significant difference in the short wavelength region < 550 nm. With further increase in wavelength, quantum efficiency falls to 0% for $Cu_2O/TiO_2$ in 570 nm. For $CuO/TiO_2$, it is nearly constant to 820 nm and then drops to 0%.

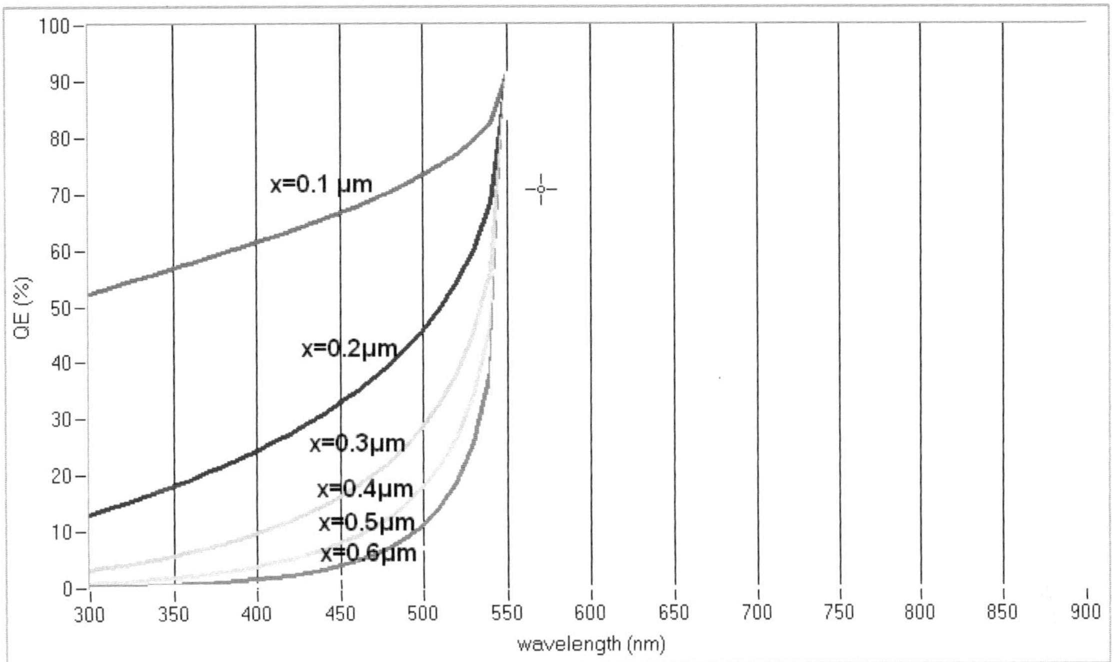


**Figure 5a.** Quantum efficiency (QE) of the $Cu_2O/TiO_2$ solar cell at different layers thickness of $TiO_2$.

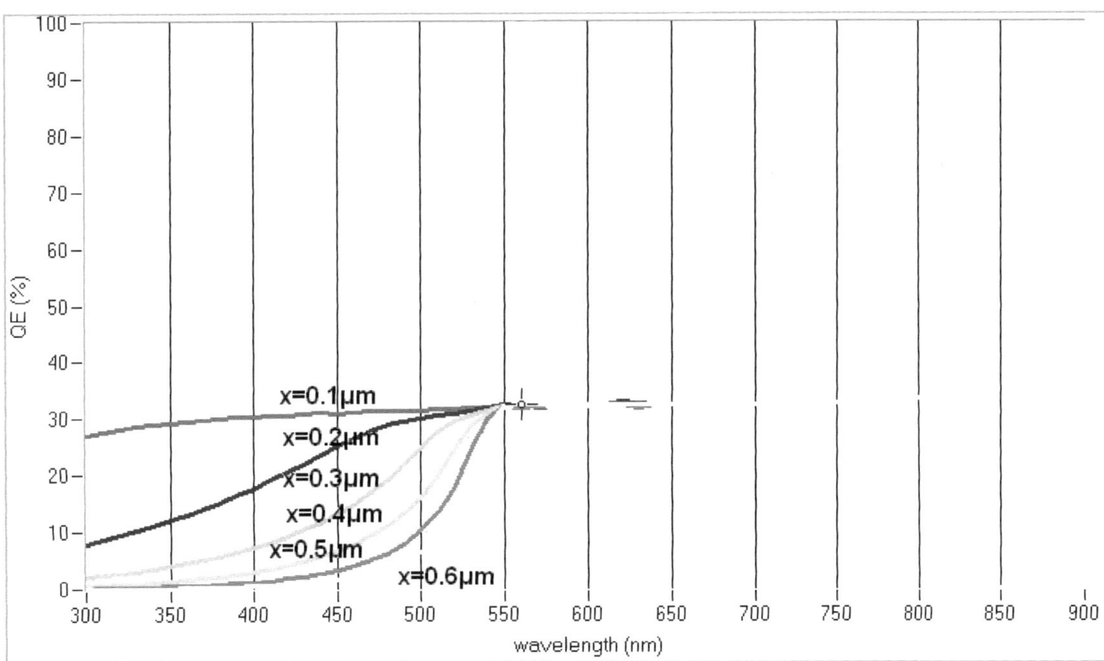


**Figure 5b.** Quantum efficiency (QE) of the $CuO/TiO_2$ solar cell at different layers thickness of $TiO_2$.

The optical band gap ($E_g$) can be calculated using the model described in Ref. [57]. The calculated values and results obtain with the earlier reported works are shown in Table 3.

**Table 3.** Values of band gap for CuO and $Cu_2O$.

| | Calculated values [eV] | Literature results [eV] |
|---|---|---|
| CuO | 1.40 | 1.51 [8, 29, 49, 53, 54] |
| $Cu_2O$ | 2.38 | 2.17 [25, 41-45] |

## 6. Impact of defect density in $Cu_2O/TiO_2$ and $CuO/TiO_2$ thin film solar cell

The junction and interface layers quality are very important to cell performance, because defects can debase the quality and cause much recombination. Due to the difficulty in evaluation the recombination caused by the interface defects, two hypothetical defect layers are inserted to study the interface defects [58]. These defects could affect the structural, electronic, magnetic and optical properties [38]. In the simulation, the defects densities of n-$TiO_2$ and for p-CuO and p-$Cu_2O$ were controlled in the range 1E+12 to 1E+22 $cm^{-3}$. Fig. 6 shows efficiency (η) with different defect densities of n-$TiO_2$, p-$Cu_2O$ and p-CuO layer.

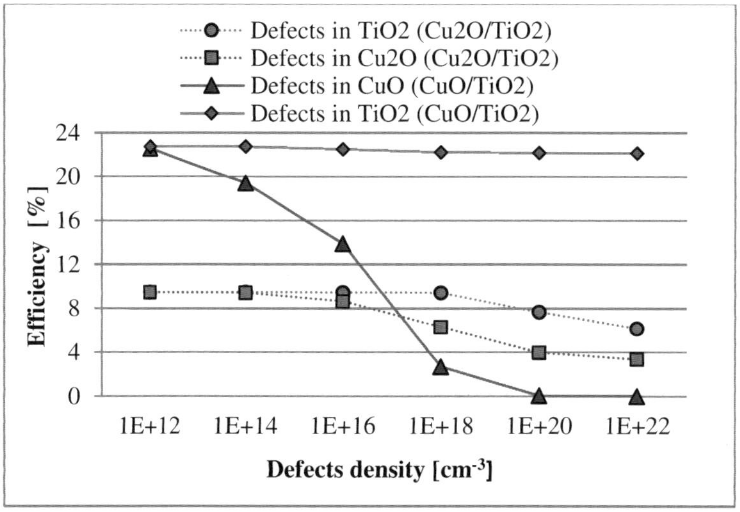


**Figure 6.** Efficiency in $Cu_2O/TiO_2$, $CuO/TiO_2$ solar cell as a function of defects density in: p-$Cu_2O$, n-$TiO_2$ layer, p-CuO layer.

The higher defect densities bring more traps and recombination centers, and deteriorate the performance of cells [58]. We note that the defect density of p-CuO layer has the most prominent influence on the device performance.

The obtained simulations reveal that for the defect density of layers <1E+18 (n-$TiO_2$) and <1E+16 (p-$Cu_2O$) the efficiency and the short circuit current remain nearly constant. With further increase of the defects density the cell parameters drop abruptly. For p-CuO layer and efficiency (η) decrease approximately linearly from 1E+12 to 1E+20.

Therefore, it is a better choice to improve $Cu_2O/TiO_2$ and $CuO/TiO_2$ interface quality by reducing defect density. The results are in agreement with the earlier reported work [58, 59].

## 7. Conclusion

In this investigation, the $Cu_2O/TiO_2$ and $CuO/TiO_2$ structures were studied using the solar cell simulator, SCAPS. By the analysis of the literature a complex simulation models of both structures were completed. The solar cells were optimized for different parameters like thickness, external quantum efficiency (EQE) of buffer layer and defect density.

On the simulation basis we have demonstrated that the variation in the thickness of the absorber layer affect strongly the conversion efficiency. The photovoltaic parameters have been calculated and it was found that optimized value of the cell thickness is 4.0 μm for p-$Cu_2O$ and CuO layers and 0.3 μm for n-$TiO_2$ layer.

For the defect density, we can also say that have a significant impact on the performance of $Cu_2O/TiO_2$ and $CuO/TiO_2$ solar cells. High defect density leads to pronounced decrease in the photovoltaic parameters, especially in p-CuO layer. It confirmed a high potential of both structures optimization with efficiencies of ~9% for $Cu_2O/TiO_2$ and ~23% for $CuO/TiO_2$ structure. Acquired knowledge will be practically utilized in the technology experiments, leading to the optimization cell construction.

**Acknowledgements**
We gratefully acknowledge to Dr. Marc Burgelman, University of Gent, Belgium, for providing the SCAPS simulation software.

**References**

[1] Li Z, Tong K, Shi R, Shen Y, Zhang Y, Yao Z, Fan J, Thwaites M, Shao G 2017 *Journal of Alloys and Compounds* **695** 3116-3123.

[2] Lu L, Guo M, Thornley S, Han X, Hu J H, Thwaites M J, Shao G, 2016 *Sol. Energy Mater. Sol. Cells* **149** 310-319.

[3] Shen Y, Guo M, Xia X, Shao G 2015 *Acta Materiala* **85** 122-131.

[4] Zou X W, Fan H Q, Tian Y M, Zhang M G, Yan X Y 2015 *RSC Adv.* **5** 23401–23049.

[5] Sibiński M, Znajdek K 2011 Innovative elastic thin-film solar cell structures *Solar Cells - Thin-Film Technologies (InTech)* 253-274.

[6] Boudaoud L, Khelifi S, Mostefaoui M, Rouabhia A K, Sahouane N 2015 *Energy Procedia* **74** 745-751.

[7] Le Zhu Development of Metal Oxide Solar Cells through Numerical Modelling, PhD thesis.

[8] Minami T, Nishi Y, Miyata T, Nomoto J I 2011 *Applied Physics Express* **4**.

[9] Li D, Chien C, Deora S, Chang P, Moulin E, Lu J 2011 *Chemical Physics Letters* **501** 446-450.

[10] Hussain S, Cao C, Usman Z, Chen Z, Nabi G, Khan W S, Ali Z, Butt F K, Mahmood T 2012 Thin Solid Films **522** 430–434.

[11] Zainun A R, Sakamoto T, Noor U M, Rusop M, Ichimura M 2012 *Materials Letters* **66** 254-256.

[12] Ichimura M, Kato Y 2013 *Materials Science in Semiconductor Processing* **16** 1538-1541.

[13] Akimoto K, Ishizuka S, Yanagita M, Nawa Y, Paul G K, Sakurai T 2006 *Solar Energy* **80** 715-722.

[14] Luo Y, Wang L, Zou Y, Sheng X, Chang L, Yang D 2012 *Electrochemical and Solid-State Letters* **15** H34–H36.

[15] Li D, Chien C, Deora S, Chang P, Moulin E, Lu J 2011 *Chemical Physics Letters* **501** 446-450.

[16] Zainun A R, Sakamoto T, Noor U M, Rusop M, Ichimura M 2012 *Materials Letters* **66** 254-256.

[17] Decock K, Khelifi S, Burgelman M 2011 *Thin Solid Films* **519** 7481-7484.

[18] Froitzheim A, Stangl R, Kriegel M, Elstner L, Fuhs W 2003 *Proceedings of the 3rd World conference on Photovoltaic Solar Energy Conversion*, Osaka (J) 1P-D3-34.

[19] Developed at Pennsylvania State University under the direction of S.J.Fonash, see: http://www.ampsmodeling.org/default.htm (last accessed January 2017).

[20] Smole F, Krč J, Furlan J 1994 *Sol. Eergy Mater. Sol. Cells* **34** 385.

[21] Burgelman M, Nollet P, Degrave S 2000 *Thin Solid Films* **527** 361-362.

[22] Frisk C, Ericson T, Li S Y, Szaniawski P, Olsson J, Platzer-Bjorkman C 2016 *Sol. Energy Mat. Sol. Cells* **144** 364-370.

[23] Zhao W, Zhou W, Miao X 2012 Numerical simulation of CZTS thin film solar cell *7th Annual IEEE International Conference on Nano/micro Engineered and Molecular Systems* 502-505.

[24] Burgelman M, Nollet P, Degrave S 2000 *Thin Solid Films* **361-362(21)** 527–532.

[25] Hossain M I, Alharbi F H, Tabet N 2015 *Solar Energy* **120** 370-380.

[26] Decock K, Zabierowski P and Burgelman M 2012 *Journal of Applied Physics* **111**.

[27] Burgelman M, Nollet P, Degrave S 2000 *Thin Solid Films* **361-362(21)** 527–532.

[28] Semiconductor device modeling with SCAPS-1D Lecture 4 Special Topics: Device Modeling, see: https://www.utoledo.edu/med/depts/radther/pdf/SCAPS-1D%20Lecture%204%20-%20handout.pdf.

[29] Serin N, Serin T, Horzum S and Celik Y 2005 *Semiconductor Science and Technology* **20** 398-401.

[30] Amin G 2012 ZnO and CuO nanostructures: low temperature growth, characterization, their optoelectronic and sensing applications *(1st ed. LiU-Tryck: Norrköping (Sweden))* [SE-601 74].

[31] Mohd Rafie Johan, Mohd Shahadan Mohd Suan, Nor Liza Hawari, Hee Ay Ching 2011 *Int. J. Electrochem. Sci.* **6** 6094-6104.

[32] Abdu Y and Musa A O 2009 *Bayero Journal of Pure and Applied Sciences* **2(2)** 8-12.

[33] Paulina Sawicka-Chudy, Grzegorz Wisz, Szymon Górny, Łukasz Głowa, Maciej Sibiński, Marian Cholewa Numerical simulation and analysis of experimental $TiO_2$/$Cu_2O$ thin film for photovoltaic structures *Journal of Nanoelectronics and Optoelectronics*

[34] Jeong S S, Mittiga A, Salza E, Masci A, Passerini S 2008 *Electrochim. Acta* **53** 2226-2231.

[35] Li B S, Akimoto K, Shen A 2009 *J. Cryst. Growth* **311** 1102-1105.

[36] Lee Y S, Heo J, Winkler M T, Siah S C, Kim S B, Gordon R G, Buonassisi T 2013 *J. Mater. Chem.* **A 1** 15416.

[37] Hossain M I, Alharbi F H 2013 *Mater. Technol.: Adv. Perform. Mater.* 28 88-97.

[38] Rivera R, Stashans A 2013 Defects in $TiO_2$ Crystals *Proceedings of the International MultiConference of Engineers and Computer Scientists (Vol II, IMECS, March 13 - 15, 2013, Hong Kong).*

[39] Kars I, Cetin S S, Kınacı B, Sarıkavak B, Bengi A, Altuntas H, Ozturk M K and Ozcelik S 2010 *Surf. Interface Anal.* **42** 1247-1251.

[40] Xosrovashvilia G, Gorjib N E 2013 *J. Mod. Opt* **60** 936–940.

[41] Gou L, Murphy C J 2003 *Nano Lett.* **3** 231-234.

[42] Wang W Z, Wang G H, Wang X S, Zhan Y J, Liu Y K , Zheng C L 2002 *Adv. Mater.* **14** 67-69.

[43] Liao L, Yan B, Hao Y F, Xing G Z, Liu J P, Zhao B C, Shen Z X, Wu T, Wang L, Thong J T, Li C M, Huang W, Yu T 2009 *Appl. Phys. Lett.* **94**.

[44] Scanlon D O, Morgan B J, Watson G W 2009 *J. Chem. Phys.* **131**.

[45] Hossain M I, Alharbi F H 2013 *Mater. Technol.: Adv. Perform. Mater.* **28** 88-97.

[46] Zhu L, Shao G, Luo J K 2011 *Semicond. Sci. Technol.* **26**.

[47] Korzhavyi P A, Johansson B 2011 Literature review on the properties of cuprous oxide $Cu_2O$ and the process of copper oxidation *Technical Report* TR-11-08.

[48] Matsumura H, Fujii A, Kitatani T 1996 *Jpn. J. Appl. Phys.* **35**.

[49] Paulina Sawicka-Chudy, Maciej Sibiński, Grzegorz Wisz Elżbieta Rybak-Wilusz, Marian Cholewa 2017 Numerical Analysis and Optimization of $Cu_2O/TiO_2$, $CuO/TiO_2$, Nanostructures PV using SCAPS *Proceedings International Conference on Microtechnology and Thermal Problems in Electronics MICROTHERM* 71-72.

[50] Lee S H, Yun S J, Lim J W 2013 The Characteristics of $Cu_2O$ Thin Films Deposited Using RF-Magnetron Sputtering Method with Nitrogen-Ambient *ETRI J.* **35**.

[51] Ritter E 1975 *Phys. Thin Films* **8**.

[52] Liangbin Xionga, Sheng Huanga, Xi Yanga, Mingqiang Qiuc, Zhenghua Chena, Ying Yua, 2011 *Electrochimica Acta* **56(6)** 2735-2739.

[53] Jun-Yong Parka, Chan-Soo Kimb, Kikuo Okuyamac, Hye-Moon Leed, Hee-Dong Jange, Sung-Eun Leef, Tae-Oh Kima 2016 *Journal of Power Sources* **306** 764-771.

[54] Masaya Ichimura, Yoshihito Kato 2013 *Materials Science in Semiconductor Processing* **16(6)** 1538-1541.

[55] Zhao W, Zhou W, Miao X 2012 Numerical simulation of CZTS thin film solar cell *7th Annual IEEE International Conference on Nano/micro Engineered and Molecular Systems* 502-505.

[56] Liao L, Yan B, Hao Y F, Xing G Z, Liu J P, Zhao B C, Shen Z X, Wu T, Wang L, Thong J T, Li C M, Huang W, Yu T 2009 *Appl. Phys. Lett.* **94**.

[57] Koffyberg FP and Benko FA, 1982 *Journal of Applied Physics*.

[58] Kai Tan, Peng Lin, Gang Wang, Yan Liu, Zongchang Xu, Yixin Lin 2016 *Solid-State Electronics* **126** 75-80.

[59] Mostefaouia M, Mazaria H, Khelifia S, Bouraioua A, Daboua R 2015 *Energy Procedia* **74** 736-744.

IOP Conf. Series: Journal of Physics: Conf. Series **1033** (2018) 012003 doi:10.1088/1742-6596/1033/1/012003

# Power electronic supply system with the wind turbine dedicated for average power receivers

**Tomasz Widerski**[1], **Adam Skrzypek**[2]

[1] Lodz University of Technology, Wolczanska 211/215, 90-924 Lodz, Poland; tomasz.widerski@p.lodz.pl
[2] ABB Sp. z O.O., Placydowska 27, 95-070 Aleksandrow Lodzki, Poland

E-mail: tomasz.widerski@p.lodz.pl

**Abstract.** This article presents the original project of the AC-DC-AC converter dedicated to low power wind turbines. Such a set can be a good solution for powering isolated objects that do not have access to the power grid, for example isolated houses, mountain lodges or forester's lodges, where they can replace expensive diesel engine generators. An additional source of energy in the form of a mini-wind farm is also a good alternative to yachts, marinas and tent sites, which are characterized by relatively low power consumption. This article presents a designed low power wind converter that is dedicated to these applications. The main design idea of the authors was to create a device that converts the very wide range input voltage directly to a stable 230VAC output voltage without the battery buffer. Authors focused on maximum safety of using and service. The converter contains the thermal protection, short-circuit protection and overvoltage protection. The components have been selected in such a way as to ensure that the device functions as efficiently as possible.

## 1. Power electronics solutions in wind energy area

### *1.1. The wind energy development*

Based on current data and forecasts presented by experts, significant increases in energy prices from fossil fuels are expected. Developed countries have long recognized the problem and started intensive progress of alternative energy. Germany is the perfect example, where in 2014 the largest share of renewable energy production took place. Currently, wind turbines are the largest source of renewable energy. Over the last twenty years, there has been a noticeable rise of wind energy area in the world. The increase in the capacity of these power plants in the last twenty years was more than 9000% [1,2,3,4]. This phenomenon is best illustrated by the fact that the combined capacity of wind turbines installed in Poland alone in 2015 was higher than in 1995 worldwide [5,6]. Trends in the development of global wind power are illustrated in figure 1 [1,2]. The power growth of the installed wind power plant is about 20-30% per year and it can be assumed that this trend will continue.

Due to the huge interest in the generation of energy from wind farms, the production of components used for their construction is growing steadily and the number of companies involved in the development of this energy sector is growing. As a result, the cost of wind farms is reduced, which further accelerates the development of this sector.

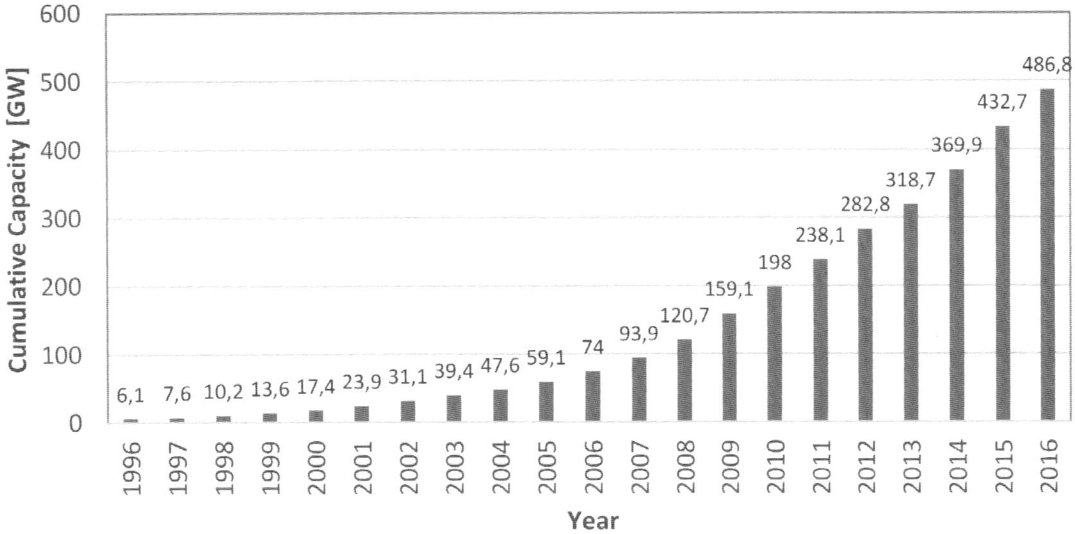


**Figure 1.** Global wind power cumulative capacity. [source: GWEC]

It should also be noted that the generation of wind energy is increasingly available to individual people and small companies with limited budget. On the market there are companies producing small power plants below 30kW, which can be purchased at an attractive price. This allows to construct of a small energy generator at the place where it is consumed. This is an attractive solution, due to the significant reduction of transmission losses. The advantages of using low power wind turbines are most evident in the supply of isolated non-grid facilities, such as lonely houses, mountain huts or forester's houses, where they can replace expensive diesel engine generators. An additional source of energy in the form of a mini-wind farm is also a good alternative to yachts, marinas and tent sites, which are characterized by relatively low power consumption.

### *1.2. Converters for high power generators*

One of the biggest problems of wind energy is the variability of the generators output voltage electrical parameters. As the wind speed changes, the rotation speed of the turbine is changed and therefore the output voltage and amplitude are changed. To solve this problem, AC/DC converters are installing in wind generator plants to set the stability output voltage value, suitable for the energy consumers. These are switching circuits containing IGBT transistors or IGCT thyristors [7] (low power converters contain MOSFET transistors). An example of such a converter for handling large wind turbines is the PCS 6000 model developed by ABB company [8], whose block diagram is shown in figure 2. Depending on the configuration, the converted power can be up to 9MVA at 3.3kV output voltage and 98% efficiency.

The voltage generated in the generator (a) is converted by the thyristor rectifier (d). Between these blocks there is a necessary disconnector (b) to disconnect the generator from the converter for e.g. service the device. The filter (c) protects the generator from the hazardous overvoltages generated during the switching of the rectifier power diodes. High amplitude overvoltages can damage the insulation in the generator windings and lead to the failure of the entire power plant. The middle section (e) is responsible for eliminating the ripple voltage in the rectifier. Also included in this block is a converter discharging circuit made up of two IGCT thyristors. Another converter section is the inverter (f), which is not different from the rectifier. Its purpose is to generate a three-phase voltage with a sinusoidal waveform. In order to improve the quality of the output voltage, an LC low pass filter (g) is used to eliminate the higher order harmonics. The last element of the circuit is a three-phase transformer that transforms the average voltage to the mains voltage (often 15kV) and the grid disconnector (h).

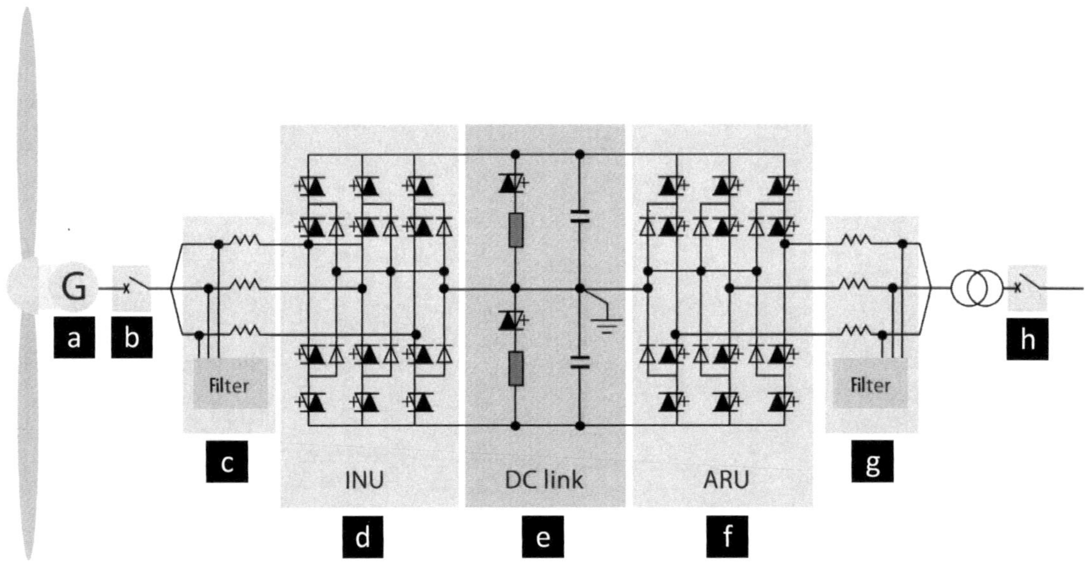


**Figure 2.** PSC 6000 WIND converter topology (a – generator, b – generator disconnector, c – input filter, d – rectifier, e – capacitors bank and discharging system, f – inverter, g – output filter, h – output disconnector) [8].

*1.3. Converters for low power generators*

There are relatively few commercial solutions for small wind converters for low power wind turbines. An example is AirGenerator company, which offers wind converters from 1kW to 20 kW, 230VAC / 50Hz output voltage and 80% efficiency (depending on output power) [9].

The above mentioned wind converters are characterized by high stability of the output voltage parameters. Unfortunately, they require high-stable input voltage, hence the wind generator and converter cannot be directly connected. In this case, a rechargeable battery is used, whose role is to receive electricity from the generator when the output power is higher than that received. Consequently, the converter contains a battery charging regulator, whose purpose is to stabilize the voltage generated by the wind turbine. This problem is typical for low input voltage converters. For a 1 kW model, the supply voltage is 48VDC, which means high input currents, significant power losses during energy converting and, as a result, decrease efficiency. The efficiency of inverters increases considerably for higher power and higher input voltages [9].

Low power converters are significantly different from the high power converters designed to wind power plants. IGBT transistors are used in low power applications. They operate at frequencies from 1kHz to 40kHz (depending on the model) [10]. Increasing the switching frequency allows the generation of sinusoidal waveforms with small deformations. The output filter of this converter may be smaller and therefore cheaper. The converter also generates less noise.

The high power converter, as mentioned above PCS 6000 WIND, has a longer switching time than IGBT transistors, so the switching frequency is less than 1kHz for output 5kV voltage and 2.5kA current. This enables the reduction of the power losses and thus high efficiency. On the other hand, the output voltage is significantly deformed, which requires the use of a high-inductance filter , which generates large electromagnetic interferences. In addition, due to high losses (over 100kW for full output power), liquid cooling of the switching devices and chokes is used. This solution complicates the mechanical construction (external radiator), electrical scheme (additional sensors and cooling system controls) and increases costs. The control circuit of the converter controls many subsystems and sensors, which is extremely complicated compared to low power converters. However, this is necessary because of the risk of fire or breakdown.

IOP Conf. Series: Journal of Physics: Conf. Series **1033** (2018) 012003 doi:10.1088/1742-6596/1033/1/012003

## 2. Project of low power wind converter

Based on the information presented above, the authors have developed a converter designed to convert energy from an unstable source such as a small wind generator. The main authors design idea was to create a device that converts the very wide range input voltage directly to a stable 230VAC output voltage without the battery buffer. Most commercial wind power converters include battery charging systems or are dedicated to connect to the power grid [11,12,13,14,15,16]. There are few solutions that would allow to supply more complex facilities (several receivers) and at the same time were relatively cheap and without charging systems or phase synchronization. The solution is to use dedicated converters for special applications, such as powering the receivers on the yacht [17,18], but they are usually very expensive, because they must meet very high criteria for resistance to weather conditions or safety. In addition, the solutions on the market usually have a rather high input voltage level at which the converter starts works (usually 100-150V or higher) [13,19]. An additional problem is the fact that the manufacturers limit the universality of the proposed converters. Usually, they are dedicated to wind turbines offered by a given manufacturer. The converter designed by the authors has a very wide range of input voltages, thanks to which it gives the output voltage even at low input voltages, and is able to cooperate with any type of wind generator, which makes the solution extremely versatile.

### *2.1. General design assumptions*

The converter design is based on the following assumptions:

- Device safety (the protection of user, external environment and internal components of the converter; the overvoltage protection from the high voltage from generator and the output short-circuit protection).
- Use of three-phase rectifier in the input. (possibility of cooperation with different wind generators).
- Input voltage from 0 to 330V (after rectifier).
- Output voltage 230VAC.
- Continuous output power at 1kW.
- Auxiliary 12V power supply for switching transistors drivers (galvanic isolating power supply).
- Engineering solutions that reduce the price of the device and increase its reliability.
- Achieving the highest possible performance.
- Quiet operation and relatively small dimensions.
- Simple interface for easy operation by serviceman.

The converter block diagram, including the power switches and the most important passive components, is shown in figure 3. The power of the converter is divided into three parts: rectifier, voltage converter and inverter together with the output filter.

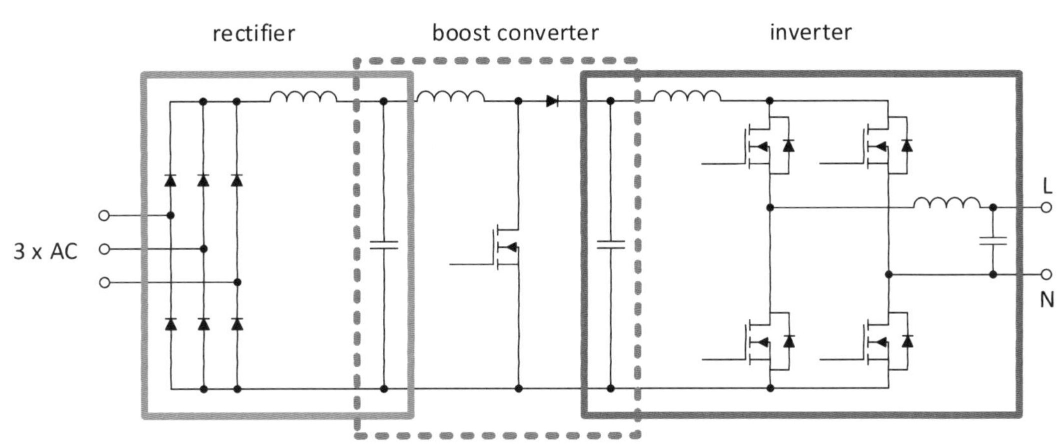


**Figure 3.** The electrical diagram of the designed wind converter.

The rectifier section has three input connectors for single-phase or three-phase voltage. The overcurrent protection is located in this section too. The aim of this one is to protect the generator against damage in the event of a converter malfunction. At the output of this section there is a bank of capacitors filtering the voltage and stabilizing the parameters of the converter. The choke between the rectifier bridge and the capacitors reduces the ripple of the current flowing between them, thus increasing the life span of the wind turbine and reducing the electromagnetic interference generation.

The next section is a boost converter that increases the input voltage to about 325V. The value of this voltage allows inverter generate a sinusoidal waveform with an effective voltage close to 230VAC. The driver controlling the operation of the switching transistor measures the output voltage and sets the value of PWM coefficient of gate impulse. The frequency of the MOSFET switch is about 70kHz. Another task of the system is to control the current value drawing from the wind turbine, which makes it possible to use the generator optimally and to prevent it from overheating. This section also includes the converter unloading switch when the unit turns off. The energy accumulated in the capacitors is decomposed on a resistor connected in series to the discharge transistor.

The last section of the wind converter is a inverter, which convert DC voltage for AC voltage (sinusoidal). For the output waveform to have a sinusoidal shape, the switching transistors should work at a frequency of 20kHz. The input choke of the inverter reduce the switching losses of the power transistor and reduce the electromagnetic interference that occurs when operating with heavy loads. At the output of the inverter there is a LC filter modifying the rectangular waveform generated by the inverter to a sinusoidal waveform. The higher the inductance value of the choke and capacitor capacitance, the more accurate the sinusoidal waveform, but also the greater the voltage drop across the filter (higher heat loss and lower output voltage).

### *2.2. Block diagram of the designed converter*

Due to the large size of some components, a few blocks of the wind converter are divided into smaller modules, such as the placement of the inverter and the filter along with the output circuit on the various PCB board. A block diagram illustrating the connections between the modules is shown in figure 4. The schematic contains also an isolated power supply and a main microcontroller based on the Atmega16A. The task of the microcontroller is:

- Control of relays placed in front of the rectifier (disconnecting the generator and converter if the input voltage is too high).
- DC voltage measurement before and after the inverter.
- Starting up the boost converter and limiting the maximum current through it.
- Temperature measurement of switching elements located in the boost converter and inverter. This solution allows the control of the switching devices cooling fan and protect the system from overheating.
- Control of inverter power transistors.
- Control of capacitors battery discharge transistor.
- Starting the isolated power supply.
- Control of relays located on the output of the wind converter (converter and load disconnection).
- Measurement of the output voltage of the converter.
- Output current measurement (short circuit detection on load).

An isolated power supply is needed to ensure proper control of the inverter transistors. In addition, this power supply supplies the bank of the filter capacitors and the boost converter. It is designed as a set of DC/DC converters. Galvanic isolation provides small impulse transformers. Due to the low power consumed by the controllers and the efficiency of the power supply over 80%, it does not introduce significant power losses to the entire wind converter.

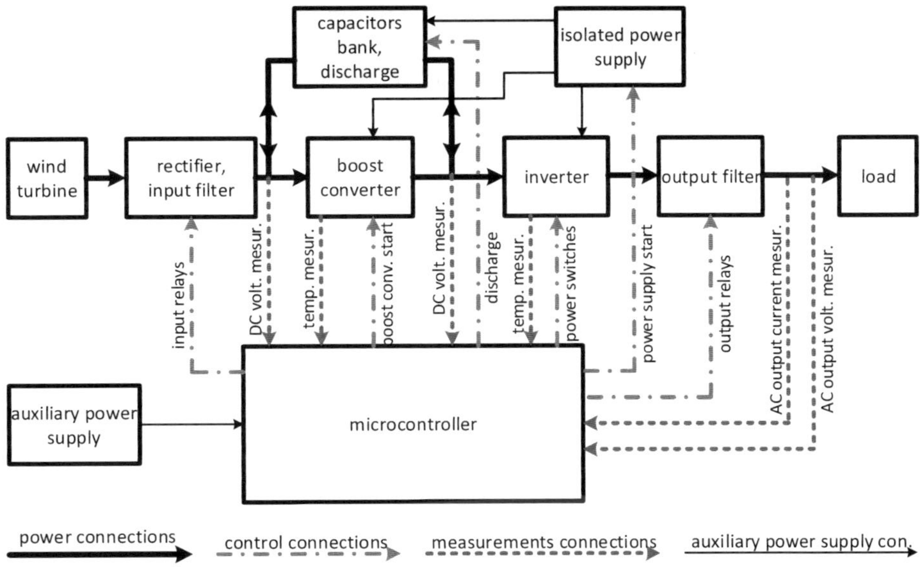


**Figure 4**. The block diagram of the wind converter.

Between the generator and the rectifier there is additionally short circuit protection in the form of a three-phase overcurrent turn-off switch with a maximum current of 6A. Also between the output circuit of wind converter and the load is a single phase turn-off switch with a maximum current of 6A. The main capacitors are discharged via the earthing switch.

The control system has the ability to receive signals from the sensors that signal the opening of the converter case and to activate the main capacitor discharge circuits, which significantly reduces the risk of electric shock. To enable communication with the converter driver, the project provides an operator interface consisting of:

- Power switch for control circuit,
- Power switch for other modules,
- Converter reset buton,
- Main converter switch,
- Indicator light for the control system supply,
- Indicator light for other modules power supply,
- Warning indicator light for low level of inverter input voltage,
- Warning indicator light for earthing connected and disconnecting converter from the generator or load,
- Indicator lamp for correct operation of wind converter (capacitors voltage above 30V, converter connected to generator and loads, inverter and boost converter power transistors working).
- Warning light,
- Safety switch,
- An ammeter measuring the current taken from the generator (behind the rectifier).

*2.2.1. Boost converter*

One of the most important converter circuits is the boost converter. Due to the selected topology presented in figure 5 [10,20,21,22] and the nature of the work, it is the module with the lowest efficiency. This is effect of unfavorable switching conditions of the power semiconductor (high voltage and current) and voltage drops on the inductance and diode. The alternative is to use a transformer converter with higher efficiency, but in this case the converter may not directly work with unstable input voltages like wind turbines (without battery buffer) [10,22,23]. For this reason, a simpler topology of the boost converter has been used at the expense of lower efficiency, which is evident at low input voltages. The converter works in CCM mode (Continuous Current Mode) [22].

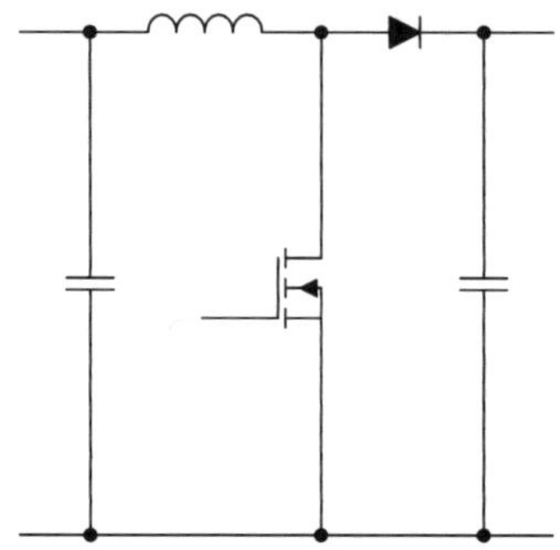

**Figure 5.** The electrical diagram of the boost converter.

The MOSFET transistor (IRFP460) is switching at 70 kHz and PWM from 10% to 97%. The power lost in the transistor, for PWM = 97%, is 10.77 W. The boost converter control circuit includes a short-circuit protection and a feedback circuit that controls the output voltage. During testing of the described block with a resistive load of 800 W, the correct output voltage (Uout = 329 V) was reached at the input voltage of 250 V, as shown in table 1. It should be noted that the high level of output voltage stabilization and increase of efficiency with the increase of input voltage. During the measurements, slight increase in temperature was observed on the power components (up to 45°C), which indicates the low power losses and thus the high energy efficiency of the inverter.

**Table 1.** The efficiency of the boost converter with resistance load for different input voltage.

| Vin [V] | Iin [A] | Vout [V] | Iout [A] | η [%] |
|---|---|---|---|---|
| 100.2 | 2,08 | 170.1 | 1.15 | 94.2 |
| 150.4 | 2.17 | 235.0 | 1.33 | 95.6 |
| 200.0 | 2.21 | 283.0 | 1.48 | 94.4 |
| 250.0 | 2.20 | 329.0 | 1.59 | 94.9 |
| 275.0 | 1.98 | 329.0 | 1.58 | 95.5 |
| 300.0 | 1.80 | 329.0 | 1.58 | 96.0 |

*2.2.2. Inverter*

The single phase inverter has the form of a H bridge, consisting of four power transistors IRF740, as shown in figure 6 [10,20,21]. The input and output filters of the inverter are located on a separate PCB board. The main assumptions of the inverter design include:

- Input voltage 325VDC (max 400VDC),
- Minimum output current 4A for Uout = 230VAC,
- MOSFET switching frequency inverter 20kHz,
- Galvanic isolation between the control circuit and power transistors by optocouplers.

It was important to fit the large radiators for MOSFETs. Their effectiveness largely depends on the efficiency of the inverter. The power losses, calculated for one 20kHz switched transistor, amounted to 4.65 W. Due to the increase in $R_{DSon}$ resistance along with the temperature, only a forty degree increase in the temperature of the junction was allowed, which necessitates the use of larger heat-sinks. This allows a short overload of the inverter, but increases the cost, weight and size of the device. To avoid these disadvantages, authors were decided to use a cooling fan. The power consumption of the additional fan (2 W) is disproportionately lower compared to the increase in power losses on the switching elements.

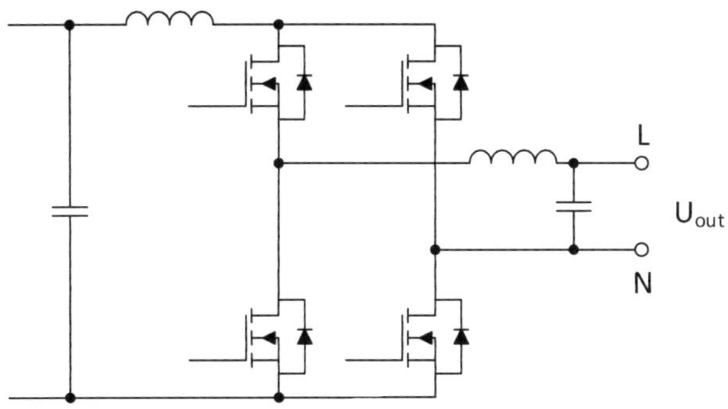


**Figure 6**. The electrical diagram of the inverter.

In order to cool down only when the inverter or boost converter is overheated, it was decided to use NTC thermistors as temperature sensors mounted on the heat sink of the boost converter transistor and on the heatsink of one of the inverter transistors. This solution reduces power losses and allows the converter to run quietly at low loads.

*2.2.3. Controlled system*

The control module is the most expanded module in the converter. It monitors the current operating parameters of the wind converter and sends the control signals to the other modules: inverter, boost converter and discharge circuit. Its block diagram is shown in figure 7. It includes the Atmega16A microcontroller and steering-measurement systems (control of output voltage, load current, power supply voltage, radiator temperature, inside and outside of the case temperature and rotation speed of the wind turbine) [24]. In addition, it controls the relays disconnecting the wind converter from the generator and the load, the cooling fan and controls the opening of the case. In order for the converter to work long enough after power failure, the system is equipped with a supercapacitor bank. All control blocks are based on simple and effective solutions [25], therefore the whole system is cheap and not very extensive.

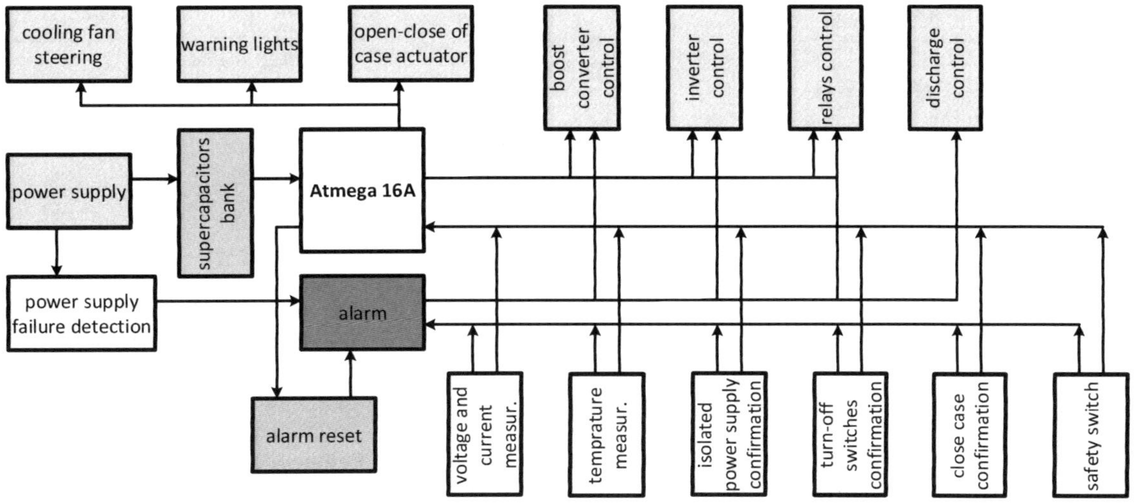


**Figure 7.** The block diagram of the control system.

IOP Conf. Series: Journal of Physics: Conf. Series **1033** (2018) 012003 doi:10.1088/1742-6596/1033/1/012003

### *2.3. Microcontroller software*

A very important part of the project is the software of the microcontroller controlling the work of the converter blocks. In the case of this wind converter project, the Atmel microcontroller with AtmelStudio 6.2 was chosen. Programming language "C" has been used in the application [26,27]. The programmer AVRISP MkII was used to upload the compiled file to the microcontroller.

Figure 8 shows a flowchart control algorithm. The first instructions in the program are the definitions of the variables and configurations of the respective microcontroller registers. The next step is to start the A/D converter and the interrupt enable bits. The last step of configuring the microcontroller is to determine the mode of operation of the individual terminals (as input or output). All these operations are performed only once after the microprocessor is started.

The program analyses the inputs of the microcontroller and converts measurements obtained from the A/D converter. On this basis, it determines if an alarm should start and the main wind converter blocks should switched off. Based on the received signals, the microcontroller controls the digital outputs.

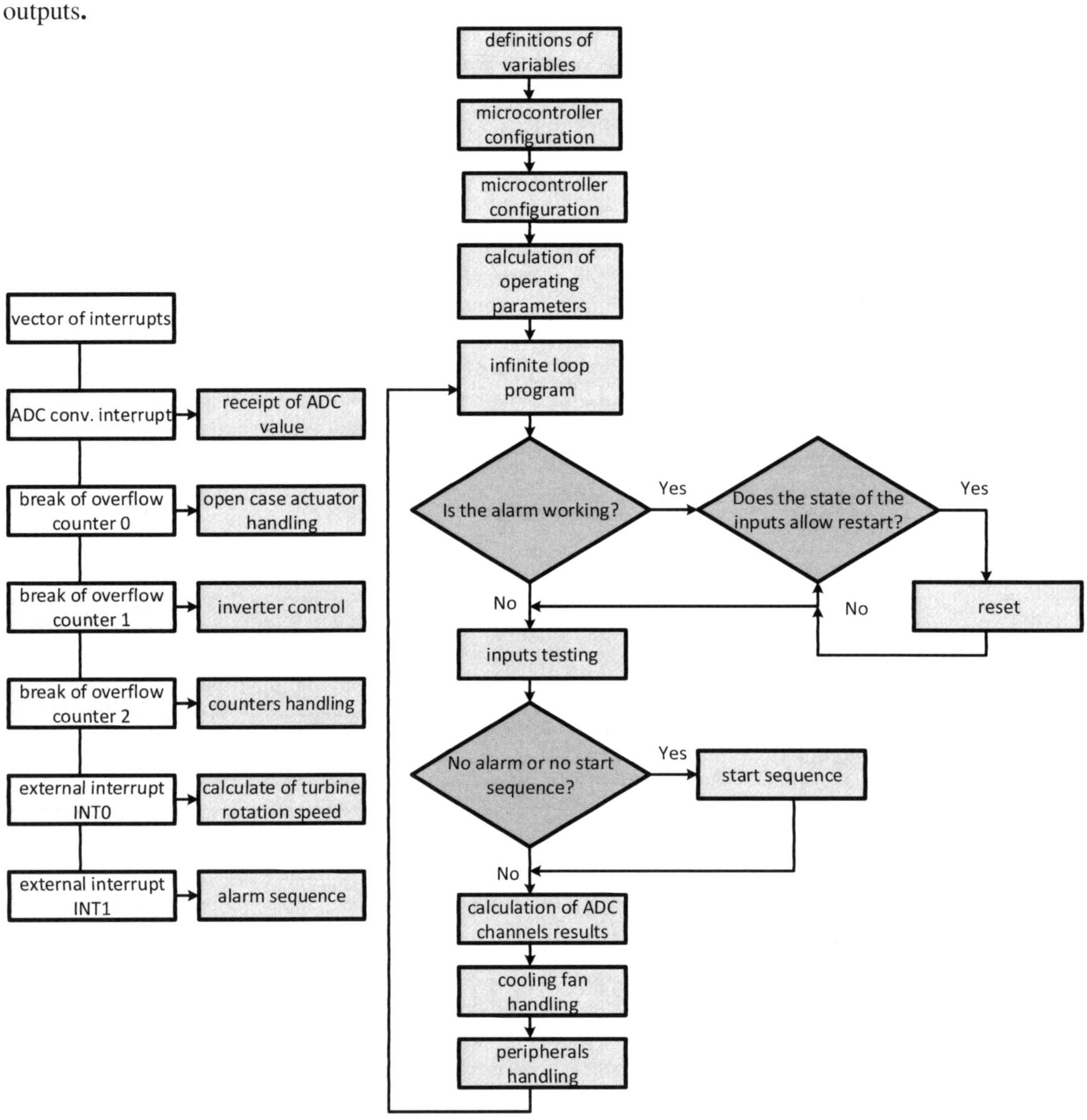


**Figure 8.** Algorithm of control program. (red - decision blocks, violet - blocks executed only during program initialization, yellow - interrupt, green - blocks executed during main loop operation).

IOP Conf. Series: Journal of Physics: Conf. Series **1033** (2018) 012003 doi:10.1088/1742-6596/1033/1/012003

## 3. Inverter tests

The main aim of the inverter test with different load is to check the thermal response of the system. Components such as inverter and boost converter power transistors or the output filter choke, produce appreciable amounts of heat during the current flow. As part of the tests, their case temperatures were measured by digital multimeter with thermocouple. The first test was performed with an inductive load 2.75 mH. The main capacitors were charged to approximately 51 V and then the inverter was started with a 1/8 sinusoidal inverter so as to achieve a rated current of 4 A. The ambient temperature was 27.5°C and the fan was off. The following temperature values were measured:

- The inverter power transistor achieve 55°C, resulting in an increase in the $R_{DSon}$ resistance and a decrease the output current to 3.7 A. Raising the voltage to 55 V increases the output current and automatically causes a further increase switch transistors temperatures and the process repeats. Activation of cooling fan immediately reduces the temperature of the inverter's heatsink, after a 20 minutes their temperature stabilizes at 30°C and the current rises to 4.3 A (at 55 V on the capacitors).
- The measurement of the choke temperature after about 10 minutes indicated stable value of 38°C.

As part of the next test, authors forced the overloaded of the system. The capacitor voltage has been increased to about 74 V using the same load and output current achieves 6 A. Due to the risk of damage to the converter, the test was carried out while the fan was running. The following temperature increases have been reported:

- The inverter transistors have reached 36°C (at an ambient temperature of 27.5°C). This confirms the effectiveness of forced cooling.
- The choke has reached 55°C after about 5 minutes, which is already a high value for this type of element (6 A rated current of the choke).

Measurements made during the tests show that the semiconductor power components are not even close to the maximum temperature values given in the catalogue notes [28,29,30]. The main limitation in this case is excessive overheating of the output filter choke, but only after 150% of the converter overload. The last test checked the electrical breakdown endurance of the inverter block. The mains capacitor voltage has been increased to 335 V and the load has been exchanged for resistance load. The output current of the wind converter was 3.42 A. The test confirmed the correct operation of the device and none of the power components showed excessive overheating. This is indicated by the temperature of the inverter transistors and the output filter choke shown in table 2. The test also checked the quality of the output circuit under load. The oscillogram of the output voltage of the wind converter is shown in figure 9.

**Table 2.** Temperatures of inverter power transistor and filter inductor for different input voltage and load (with working fan)

| | Transistor temperature | Choke temperature |
|---|---|---|
| Vin=55V Iout=4,3A | 30°C | 38°C |
| Vin=74V Iout=6A | 36°C | 55°C |

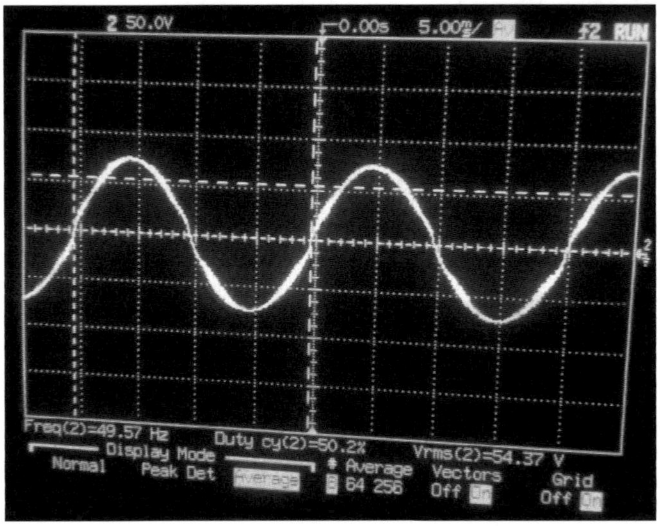


**Figure 9.** The oscillogram of the output voltage of the wind converter.

## 4. Conclusions

The purpose of this work was to design a power converter for a low power wind generator and small and medium 230VAC loads. The authors have developed switching blocks (boost converter and inverter), additional filter systems and a control system to monitor wind converter operation, prevent failures and ensure safe use and service. The whole is controlled by a microcontroller, which works based on the author's software. It is worth highlighting that the whole project also takes into consideration the aspects of the safety of the converter and the minimization of losses on the semiconductor and inductance elements. In addition, information on testing individual blocks is included.

Although the designed converter meets the initial requirements and covers the safety problems, there are several issues that can be developed to improve the quality of this wind converter. The most important modification of this one to be considered in the future is the replacement of MOSFET transistors on IGBT transistors or SiC-based components. This solution will increase the power of the inverter without changing the heat sink and cooling system. The use of IGBT switches will enable at least double the maximum power of the converter and reduce of losses. However, that it leads to the redesign the remaining circuits, especially inside the boost converter.

Another possibility of this converter development concerns the communication with PC computer by Bluetooth. The HC-06 module is optionally installed on the PCB controller board when designing the printed circuits. An application running on the computer would allow to track the current parameters of the converter working and control the individual components such as inverter or boost converter. In addition, the program would record errors occurring during the operation of the device, which would significantly facilitate detection of any malfunction and repair it.

The last potential development is design a module, which may be able to provide energy from outside power grid to the inverter power supply. Thanks to this, in the case of weak winds, the energy would be partially taken from the grid, which would allow the receivers demanding a continuous power supply. The use of such a module would greatly increase the versatility of the converter, since it would enable almost all types of receivers.

## 5. References

[1] *Renewable Energy Technologies: Cost Analysis Series, Volume 1: Power Sector, Issue 5/5 Wind Power* (IRENA Working Paper), June 2012, https://www.irena.org/documentdownloads/publications/re_technologies_cost_analysis-wind_power.pdf (access 26.03.2017)

[2] GWEC - global statistic, http://gwec.net/global-figures/graphs/ (access 26.03.2017)

[3] The European Wind Energy Association - EWEA, *Wind in power 2015: European statistics*, February 2016, https://www.ewea.org/fileadmin/files/library/publications/statistics/EWEA-Annual-Statistics-2015.pdf (access 26.03.2017)

[4] *Renewables 2016 Global Status Report* (by REN21), http://www.ren21.net/wp-content/uploads/2016/06/GSR_2016_Full_Report.pdf (access 26.03.2017)

[5] Energy Regulatory Office (Poland) – map of RES, http://www.ure.gov.pl/uremapoze/mapa.html (access 26.03.2017)

[6] *Global Wind 2006 Report* (by GWEC) http://gwec.net/wp-content/uploads/2012/06/gwec-2006_final_01.pdf (access 26.03.2017)

[7] ABB product brochure 2015 *IGCT – integrated gate-commutated thyristors*; http://www.brown.edu/Departments/Engineering/Courses/ENGN1931F/ABB_Flyer_IGCT_2015.pdf (access 27.03.2017)

[8] ABB product brochure 2016 *PCS6000 wind turbine converter Medium voltage, full power converter, up to 12 MW*; http://new.abb.com/power-converters-inverters/wind-turbines/utility-scale/pcs6000 (access 28.03.2017)

[9] *Przydomowe elektrownie wiatrowe* (by AirGenerator) http://generatory-wiatrowe.pl/produkty/elektrownie-wiatrowe/ (access 28.03.2017)

[10] Ke Ma 2015 *Power Electronics for the Next Generation Wind Turbine System* (Switzerland: Springer International Publishing)

[11] Teodorescu R, Liserre M, Rodríguez P 2011 *Grid Converters for Photovoltaic and Wind Power Systems* (John Wiley & Sons Ltd published)

[12] Fattah Hassanzadeh, Hossein Sangrody, Amin Hajizadeh, Shahrokh Akhlaghi 2017 *Back-to-back Converter Control of Grid-connected Wind Turbine to Mitigate Voltage Drop Caused by Faults*, Conference: 49th North American Power Symposium, September 2017, https://www.researchgate.net/publication/318205807_Back-to-back_Converter_Control_of_Grid-connected_Wind_Turbine_to_Mitigate_Voltage_Drop_Caused_by_Faults (access 08.03.2018)

[13] Nayar Ch, Dehbonei H, Chang L, 2005 *A Low Cost Power Electronic Interface for Small Scale Wind Generators in Single Phase Distributed Power Generation System*, (AUPEC 2005 Australasian Universities Power Engineering Conference, Australia, 25 -28 September, 2005), http://citeseerx.ist.psu.edu/viewdoc/summary?doi=10.1.1.574.1392 (access 10.03.2018)

[14] Blaabjerg, F., Iov, F., Chen, Z., & Ma, K. 2010 *Power Electronics and Controls for Wind Turbine Systems*. (In Proceedings of the IEEE Energy Conference and Exhibition, ENERGYCON 2010, Bahrain) http://vbn.aau.dk/files/46616844/Power_Electronics_and_Controls_for_Wind.pdf (access 10.03.2018)

[15] Krishnamoorthy H, Rana, D, Garg P, Enjeti PN and Pitel I, Fellow *Wind Turbine Generator–Battery Energy Storage Utility Interface Converter Topology With Medium-Frequency Transformer Link*, (IEEE TRANSACTIONS ON POWER ELECTRONICS, VOL. 29, NO. 8, AUGUST 2014), http://ieeexplore.ieee.org/stamp/stamp.jsp?tp=&arnumber=6689314 (access 10.03.2018)

[16] Mayouf M, Abdessemed R *Comparative Study of a Small Size Wind Generation System Efficiency for Battery Charging*, (SERBIAN JOURNAL OF ELECTRICAL ENGINEERING Vol. 10, No. 2, pp. 261-274, June 2013), http://www.journal.ftn.kg.ac.rs/Vol_10-2/03-Messaoud-Abdessemed.pdf (access 10.03.2018)

[17] ECOSOLAR - commercial offer, http://www.ecosolar.pl/inwertery-windy-boy-c-75_76.html (access 07.03.2018)

[18] e-MARINE SYSTEMS - manufacturer's offer, https://www.emarineinc.com/categories/Inverters (access 07.03.2018)

[19] Meere R, Ibrahim I, Ruddy J, O'Loughlin C and O'Donnell T 2016 Scaled Hardware Implementation of a Full Conversion Wind Turbine for Low Frequency AC Transmission, (Energy Procedia Volume 94, pp. 182-190, September 2016)

[20] Barlik R and Nowak M 2014 *Energoelektronika: Elementy, Podzespoły, Układy* (Warsaw: Warsaw University of Technology publisher)
[21] Nowak M and Barlik R 2013 *Poradnik inżyniera energoelektronika, tom 1* (Warsaw: WNT publisher)
[22] Schmidt-Walter H, Wenzel H, Zänker T, Morgan R and Kegan J. *Design of Switch Mode Power Supplies* http://schmidt-walter-schaltnetzteile.de/smps_e/smps_e.html (access 27.03.2017)
[23] Sanjaya Maniktala 2006 *Switching Power Supplies – A to Z,* (Burlington USA: Newnes, Elsevier Inc.)
[24] Atmega16 datasheet http://www.atmel.com/images/doc2466.pdf (access 27.03.2017)
[25] Nowak M, Barlik R and Rąbkowski J 2014 *Poradnik inżyniera energoelektronika, tom 2* (Warsaw: WNT publisher)
[26] Borkowski P 2010 *AVR&ARM7. Programowanie mikrokontrolerów dla każdego* (Gliwice: Helion publisher)
[27] Francuz T 2015 *Język C dla mikrokontrolerów AVR.* (Gliwice: Helion publisher)
[28] IRF740 datasheet http://www.vishay.com/docs/91054/91054.pdf (access 27.03.2017)
[29] IRFP460 datasheet http://www.irf.com/product-info/datasheets/data/irfp460a.pdf (access 27.03.2017)
[30] DTMSS-40 chokes datasheet http://feryster.pl/assets/nowa/pdf/dlawiki/dtmss40.pdf (access 27.03.2017)

# Investigations of the influence of frequency on power losses in ferrite cores

**Krzysztof Górecki, Kalina Detka**

Gdynia Maritime University, Department of Marine Electronics, Morska 83, Gdynia, Poland

e-mail: k.gorecki@we.am.gdynia.pl

**Abstract**. In the paper the problem of the influence of frequency on power losses in ferrite cores is considered. The method of measurements of these losses is proposed. This method is realized when direct current flows in three steps. The first of them is the measurement of the core thermal resistance in this core. Next, mutual thermal resistance between the winding and the core is measured. In the third step temperature of the inductor containing the considered core operating in a boost converter is measured at the steady state with an optical method. The power dissipated in the core is calculated using the formula elaborated by the Authors. Some results of measurements of the inductor containing ferrite cores of different dimensions are presented and discussed.

## 1. Introduction

Magnetic elements are important components of switch-mode power supplies [1]. Properties of these elements strongly depend on phenomena occurring in ferromagnetic cores [2, 3, 4]. Such cores may be made of different magnetic materials, but in the range of high frequency ferrites are particularly used [2].

The range of applications of each magnetic element determines, among others, power losses in the core. In accordance with the commonly used Steinmetz formula, power losses in the ferromagnetic core increase proportionally to frequency, which is raised with the exponent higher than 1 [2, 5, 6]. However, as it is clear from the literature [7, 8, 9, 10, 11] experimental investigations of the influence of frequency on power losses in magnetic materials are carried out only in the range of frequency not higher than several kHz. The values of power losses in the core given in the catalogue data for higher frequency are the effect of extrapolation of the measurement results using the Steinmetz formula. Higher values of frequency are taken into account in the papers [7, 12, 13, 14, 15], only.

The safe operating area of magnetic elements is limited e.g. by thermal phenomena, which cause an increase in the internal temperature of these elements. Typically, compact thermal models of magnetic devices are used [12, 13], in which uniform temperature distribution in the considered device is assumed. In such models thermal properties of the considered elements at the steady state are characterized by thermal resistance.

In the literature, e.g. [12, 16] electrothermal models of magnetic elements (inductors and transformers) that make it possible to calculate the internal temperature of the core during operation of these elements are presented. In many of the considered papers the Steinmetz formula is used to calculate power losses in ferromagnetic cores [17]. Unfortunately, this formula is elaborated for the dissipated power of the shape of a sinusoidal waveform only. Modifications of this formula are proposed e.g. in [2, 6, 13]. If magnetic elements operate in switch-mode power supplies, the waveforms of currents, voltages and power dissipated in the cores considerably differ from sinusoidal waveforms. Therefore, the problem of estimating the average value of this power occurs.

Correct modelling of properties of magnetic elements at high frequencies requires a reliable description of the influence of significant factors on power losses in the core made of magnetic materials, which can be used in devices operating at high frequencies up to several MHz. Obtaining such a description requires performance of experimental investigations, which will illustrate the influence of frequency on power losses in the core made of magnetic material [14].

The paper presents a method of measuring power losses in the ferrite core and the results of measurements of the selected pot cores of different dimensions. These results show the influence of frequency on power losses in the core used in an inductor included in the boost dc-dc converter.

In Section 2 the used measurement method is described, whereas the results of investigations are presented in Section 3.

## 2. Measurement method

The applied measuring method is an extension of the method described in [14] and it assumes that thermal properties of the core do not depend on the dissipated power in the core or frequency. Measurements are carried out in three stages, and the tested element is an inductor containing the considered ferrite core.

First, the dependence of temperature of the dissipated power in the core while feeding the core with the DC flowing through the core is measured. The value of this current is adjusted by the voltage source $V_{in}$ and the resistor R. The temperature of the core is measured with the use of a pyrometer and power losses in the core are equal to multiplication of the current flowing through it (measured by an ammeter) and the voltage on the core (measured by a voltmeter). The measurement set-up indispensable for these measurements is shown in Fig.1.

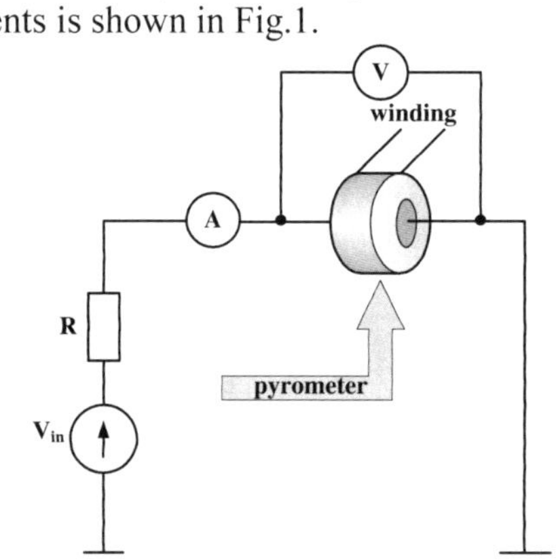


**Figure 1.** The measurement set-up used in the first step of the method.

The current value is changed in wide ranges to obtain visible changes in the core temperature as a result of a selfheating phenomenon. On the basis of the results of measurements thermal resistance of the core $R_{thR}$ is calculated and it is equal to the quotient of an increase of the core temperature $T_R$ at the steady state above the ambient temperature $T_a$ by the power dissipated in the core $p_R$.

In the second step through the winding of the inductor the direct current with the adjustable value (adjusted by the voltage source $V_{in}$ and the resistor R) flows and the core temperature is measured with the pyrometer and the voltage on the winding and the current winding are measured, too, using the voltmeter and the ammeter, respectively. The measurement set-up used in this step of measurements is shown in Fig. 2.

An increase of the core temperature over the ambient temperature is caused by mutual thermal interaction between the winding and the core. The winding current is regulated in a wide range, which causes changes in the value of the core temperature up to tens of Celsius degrees. On the basis of the results of measurements mutual thermal resistance between the windings and the core $R_{thUR}$ is calculated and it is equal to the quotient of an increase in the core temperature $T_R$ at the steady state above the ambient temperature $T_a$ by the power dissipated in the winding $p_U$. This power is equal to the product of current and voltage measured by the voltmeter and the ammeter, respectively.

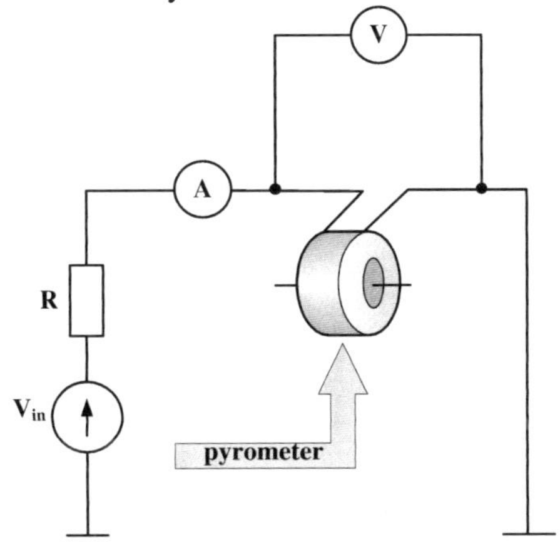


**Figure 2.** The measurement set-up used in the second step of the method.

In the third step the inductor including the investigated core and the winding used in the second step of the method is mounted in the boost dc-dc converter. During measurements, this converter should operate in the discontinuous conducting mode in order to obtain the highest maximum value of the inductor voltage. The maximum value of flux density $B_m$ in the core is proportional to this voltage. At the high value of $B_m$ a high value of the power dissipated in the core could be obtained. Through the winding the alternating current with adjustable maximum value and frequency flows. The measurement set-up used in this step is shown in Fig. 3.

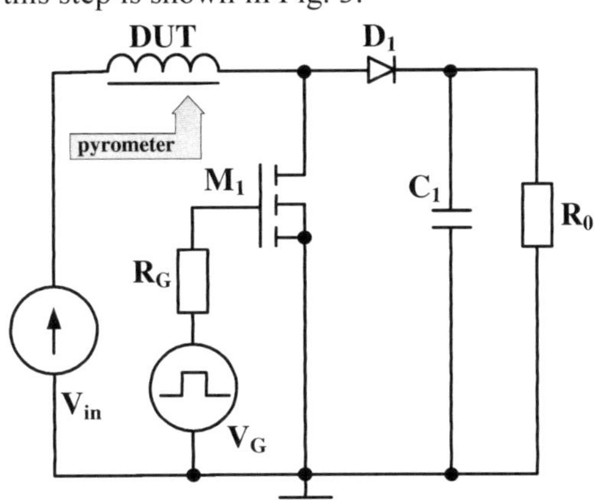


**Figure 3.** The measurement set-up used in the third step of the method.

The value of the core temperature at the steady state is measured with the pyrometer. Current and voltage of the winding are also measured. In this case, an increase of the core temperature $T_R$ is caused by selfheating in the core characterized by thermal resistance of the core $R_{thR}$ and mutual thermal coupling between the winding and the core characterized by mutual thermal resistance between the winding and the core $R_{thUR}$. A loss of power is a result of its remagnetization and by thermal coupling between the winding and the core, when in the winding the power is dissipated due to the flow of the alternating current. For each of the considered conditions of the winding supply, the average values of power losses in the core at the steady state are calculated using the formula

$$p_R = \frac{T_R - T_a - R_{thUR} \cdot p_U}{R_{thR}} \qquad (1)$$

IOP Conf. Series: Journal of Physics: Conf. Series **1033** (2018) 012004 doi:10.1088/1742-6596/1033/1/012004

where $T_a$ – the ambient temperature, $p_U$ – the average value of power losses in the core.

The proposed measurement method has some disadvantages, which limit accuracy of this method. As it was proved e.g. in [18] the thermal resistance of the core is not constant, but it is a decreasing function of the power dissipated in it. Losses in the winding caused by eddy currents are omitted. The influence of the speed of air flow and dimensions of sweet steel on thermal resistance of the core are not taken into account. Simplification assumed at the formulation of the presented measurement method does not significantly influence the accuracy of this method.

## 3. Results of investigations

The experimental investigations were carried out for inductors with two pot ferrite cores (shown in Fig. 4) of the diameter equal to 30 mm and the height equal to 19 mm (big core) and of the diameter equal to 18 mm and the height equal to 11 mm (small core) made of ferrite material F-2001. The big core was investigated for inductors with three types of windings including different numbers of turns: 54, 5 and 2. The power supply was provided to the investigated pot cores by means of connectors made of sweet steel tightened to the bases of the pot with screws. These windings are made of copper wire enamel of the diameter equal to 0.8 mm. The winding of the investigated inductor with the small core contains 2 and 5 turns of copper wire with the same diameter. Measurements of the core temperature were performed using the pyrometer Optex PT-3S with resolution of 0.1 °C.

**Figure 4**. Views of the investigated cores: a) the big core, b) and the small core.

Figure 5 shows the measured dependence of the big core (Fig. 5a) and the small core (Fig. 5b) temperature on the current flowing through the core during the first step of measurements on the power dissipated in the core during the first step of the measurement method. In Fig 5a blue and red lines denote the number of winding turns equal to 54 and 2, respectively, whereas in Fig. 5b – 2 and 5 turns, respectively.

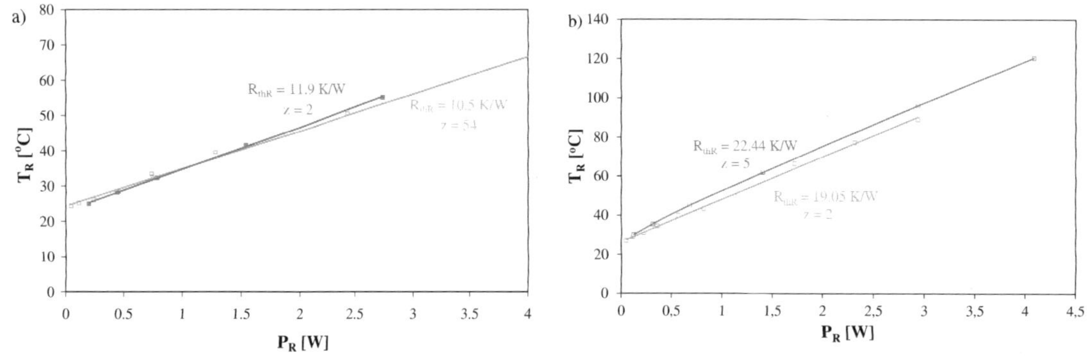


**Figure 5**. The measured dependence of the core temperature on the power dissipated in the big core (a) and the small core (b).

As can be seen, the considered dependence for both the considered cores is almost linear, which proves that thermal resistance of the core does not depend on the power dissipated in the core. It is worth noticing that the number of turns influences thermal resistance of the core $R_{thR}$. Particularly, for the small core changing this number from 2 to 5 turns causes an increase in the value of $R_{thR}$ even by 15%. Probably, the observed phenomenon is a result of worsening of the heat convection process, when the winding fills the interior of the core [19]. Such a dependence is difficult to observe for the big core. In turn, it is worth noticing that thermal resistance of the big core is twice as low as the small core. This is a result e.g. of different surface area, at which heat convection occurs.

In turn, in Fig. 6a and Fig. 6b dependences of the core temperature on the power dissipated in the winding for the big core and the small core are shown.

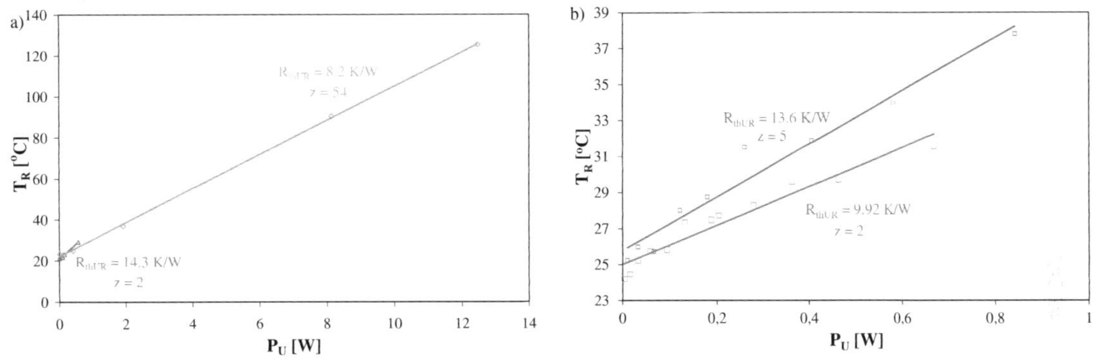


**Figure 6**. The measured dependence of the core temperature on the power dissipated in the winding of the big core (a) and the small core (b).

As can be seen, the considered dependence for both the considered cores is almost linear. The measured values of mutual thermal resistance between the windings and the core $R_{thUR}$ are smaller than thermal resistance of the core $R_{thR}$ even twice. For the big core and the small value of turns (z = 2) measurements were conducted for the power $p_U$ smaller than 0.6 W due to the small value of resistance of this winding and the limited value of the output current of the used power supply. For the small core, the strong influence of the number of turns on the value of $R_{thUR}$ is visible. In the case when z = 5 this value is even by 40% higher than in the case when z = 2.

In Figure 7 the dependence of power losses in the big core (Fig. 7a) and the small core (Fig. 7b) of the considered inductor operating in the boost converter on frequency of the control signal is presented. This converter operates at the input voltage equal to 6.75 V and load resistance equal to 1 kΩ.

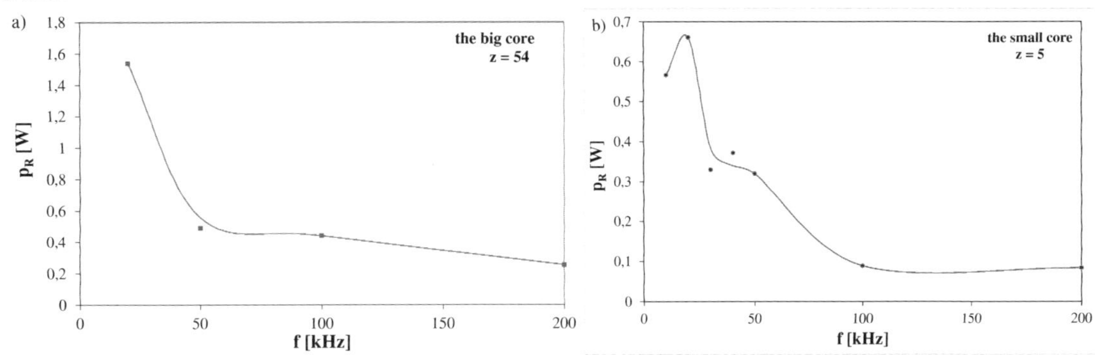


**Figure 7**. The dependence of power losses in the big core (a) and the small core (b) on frequency.

As it is visible, power losses in the core are a decreasing function of frequency. This shape of the considered dependence is a result of a decreasing dependence of the amplitude of flux density on frequency. As it is presented e.g. in the papers [6, 13, 15] the exponent in the dependence of core losses on flux density amplitude is higher than the exponent in the dependence of this power on frequency. On the basis of the obtained values of the power dissipated in the core and the measured values of $R_{thR}$ it is possible to calculate an excess of the core temperature over the ambient temperature caused by self-heating phenomena in the core. This excess can be higher than 20°C.

The power dissipated in the core of an inductor operating in the boost converter depends e.g. on frequency and the input voltage of this converter. In Fig. 8 the influence of frequency on the dependence of the power dissipated in the core on the converter input voltage is illustrated. Both the investigated inductors contain windings with 5 turns.

As it is visible, for both the considered cores the power dissipated in these cores is an increasing function of the converter input voltage. For the big core an increase of power $p_R$ while increasing frequency is observed. In turn, for the small core it is visible, that the dependence of power $p_R$ on frequency has the maximum for frequency equal to about 40 kHz. For the big core higher values of power $p_R$ were obtained.

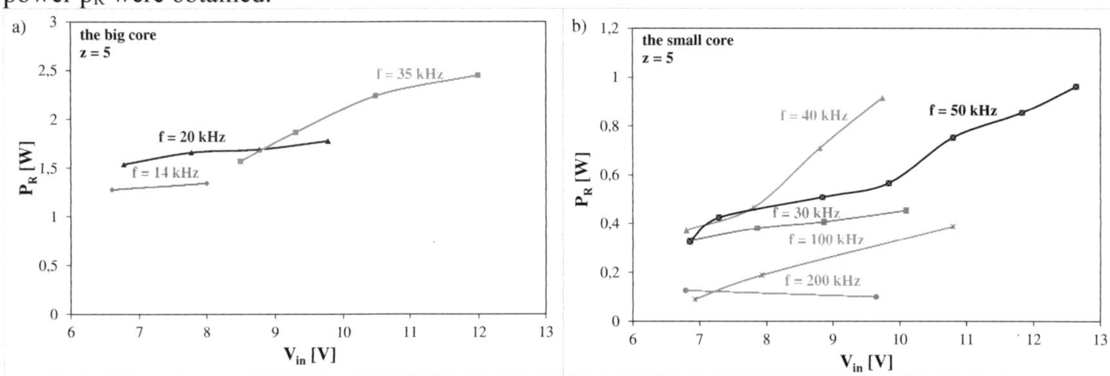


**Figure 8**. The dependence of power losses in the big core (a) and the small core (b) on the converter input voltage.

## 4. Conclusions

In the paper the method to measure power losses in ferrite cores used to build inductors is proposed. The proposed method is based on the dependence of the temperature of the core on the power dissipated in the core and the winding. This method was used to measure power losses in the selected pot cores at a wide range of frequency of the current flowing through the winding.

The obtained results of investigations prove that the power dissipated in the core included in the inductor operating in the boost converter is a decreasing function of frequency (for frequency higher than 40 kHz) and an increasing function of the input voltage of the boost converter. An increase of the core temperature caused by self-heating in the core is higher for the big core than for the small core and it can exceed even 20°C. The power $p_R$ can be high if the considered converter operates with the high value of load resistance, because in this case the converter operates in the DCM mode and the amplitude of flux density could have a high value.

The presented results could be useful for designers of switch-mode power supplies. Applications of the considered method of estimation of the average value of power dissipated in the core could make it easier estimation of the core operating temperature and verification, if the core functions in its safe operating area.

## References

[1] Ericson R, Maksimovic D 2001 Fundamentals of Power Electronics *Kluwer Academic Publisher* (Norwell)

[2] Van den Bossche A, Valchev VC 2005 Inductors and transformers for Power Electronics *CRC Press, Taylor & Francis Group* (Boca Raton)

[3] Leary AM, Ohodnicki PR, McHenry ME 2012 Soft magnetic materials in high-frequency, high-power conversion applications, *The Journal of the Minerals, Metals & Materials Society* **64** (7) 772–781.

[4] Rylko MS, Hartnett KJ, Hayes JG, Egan MG 2009 Magnetic material selection for high power high frequency inductors in DC-DC converters *Twenty-Fourth Annual IEEE Applied Power Electronics Conference and Exposition APEC 2009* doi: 10.1109/APEC.2009.4802955.

[5] Bertotti G, Fiorillo F, Soardo G 1988 The prediction of power losses in soft magnetic materials *Journal de Physique Colloques* **49** (C8) 1915-1919.

[6] Detka K, Górecki K 2017 Modelling power losses in ferromagnetic materials *Material Science Poland* **35** (2) 398-404.

[7] Sippola M, Sepponen R 2002 Accurate prediction of high-frequency power-transformer losses and temperature rise *IEEE Transactions on Power Electronics* **17** (5) 835-847.

[8] Lin D, Zhou P, Fu WN 2004 A dynamic core loss model for soft ferromagnetic and power ferrite materials in transient finite element analysis *IEEE Transactions on Magnetics* **40** (2) 1318 – 1321.

[9] Han Y, Cheung G, Li A 2012 Evaluation of magnetic materials for very high frequency power applications *IEEE Transactions on Power Electronics* **27** (1) 425-435.

[10] Muhlethaler J, Biela J, Kolar JW, Ecklebe A 2012 Improved core-loss calculation for magnetic components employed in power electronic systems *IEEE Transactions on Power Electronics* **27** (2) 964 – 973.

[11] Mu M, Li Q, Gilham DJ, Lee FC, Ngo KDT 2014 New core loss measurement method for high-frequency magnetic materials *IEEE Transactions on Power Electronics* **29** (8) 4374 – 4381.

[12] Wilson PR, Ross JN, Brown AD 2002 Simulation of magnetic component models in electric circuits including dynamic thermal effects *IEEE Transactions on Power Electronics* **17** (1) 55-65.

[13] Detka K, Górecki K, Zarębski J 2017 Modeling single inductor dc - dc converters with thermal phenomena in the inductor taken into account *IEEE Transactions on Power Electronics* **32** (9) 7025-7033.

[14] Barlik R, Nowak M, Grzejszczak P, Zdanowski M 2016 Estimation of power losses in a high-frequency planar transformer using a thermal camera *Archives of Electrical Engineering* **65** (3) 613-627.

[15] Górecki K, Rogalska M, Zarębski J, Detka K 2014 Modelling characteristics of ferromagnetic cores with the influence of temperature *Journal of Physics: Conference Series* **494** 012016 doi:10.1088/1742-6596/494/1/012016

[16] Bernardoni M, Delmonte N, Cova P, Menozzi R 2010 Thermal modeling of planar transformer for switching power converters *Microelectronics Reliability* **50** (9-11) 1778-1782.

[17] Muhlethaler J, Biela J, Kolar JW, Ecklebe A 2010 Core losses under DC bias condition based on Steinmetz parameters *Proceedings of the International Power Electronics Conference* Sapporo 2430–2437.

[18] Górecki K, Górski K, Zarębski J 2017 Investigations on the Influence of Selected Factors on Thermal Parameters of Impulse-Transformers *Informacije MIDEM – Journal of Microelectronics, Electronic Components and Materials* **47** (1) 3-13.

[19] Petkov R 1996 Optimum design of a high-power, high-frequency transformer *IEEE Transactions on Power Electronics* **11** (1) 33 – 42.

IOP Conf. Series: Journal of Physics: Conf. Series **1033** (2018) 012005 doi:10.1088/1742-6596/1033/1/012005

# Modelling nonisothermal electrical characteristics of ferrite cores

**Krzysztof Górecki, Krzysztof Górski**

Gdynia Maritime University, Department of Marine Electronics, Morska 83, Gdynia, Poland

e-mail: k.gorecki@we.am.gdynia.pl

**Abstract**. In the paper the problem of modelling an influence of thermal phenomena on electrical characteristics of ferrite cores is considered. A compact electrothermal model of the ferrite core is proposed. This model is dedicated to calculate dc and transient electrical characteristics of this core. The form of this model is presented, the method of parameters values estimation is described and some results of experimental verification of this model are shown for the planar core E22/6/16R made of material 3F3 by Ferroxcube.

## 1. Introduction

Ferrites are commonly used in electronics and power electronics circuits, in which they play a role of cores of inductors and transformers [1, 2, 3]. In these applications the most important role is played by magnetic properties of the considered materials characterised, among others by the magnetising curve B(H) [1, 4, 5]. However, during operation of magnetic elements (inductors and transformers) eddy currents are also induced in the core, which are one of the factors causing losses of energy in the core [1, 2, 6]. The value of eddy currents is inversely proportional to resistance of the core.

Ferrites belong to ceramic materials characterised with high resistivity [1]; yet, large geometrical dimensions of ferrite cores cause that this resistance does not exceed several kiloohms [7].

As it is commonly known, current-voltage characteristics of electrical elements depend on temperature [8, 9, 10, 11]. Due to self-heating phenomena the internal temperature of an electronic component is higher than the ambient temperature, and its value depends on power losses in this element [9, 10, 12, 13]. Characteristics of the electronic component calculated or measured with self-heating phenomena taken into account are called nonisothermal characteristics. In order to take into account these phenomena in computer analyses, electrothermal models of the considered elements are indispensable.

Such models are described in literature for many semiconductor devices [9, 10], inductors [3, 14] or transformers [15, 16]. For example, in the paper [8] the electrothermal model of ferromagnetic core is described, but this model takes into account the influence of thermal phenomena on magnetic properties of this core only. In the known literature there is no information about modelling an influence of thermal phenomena on electrical properties of ferrite cores.

In this paper the electrothermal model of the ferrite core making possible calculation of nonisothermal characteristics of this element is proposed, and the correctness of this model is verified experimentally for the selected ferrite planar core.

In Section 2 the form of the elaborated model is presented. In Section 3 the manner of the model

IOP Conf. Series: Journal of Physics: Conf. Series **1033** (2018) 012005 doi:10.1088/1742-6596/1033/1/012005

parameters estimation is described. Finally, the results of calculations and measurements of the core characteristics are compared in Section 4.

## 2. Model form

The presented model of the ferrite core belongs to groups of compact electrothermal models [8, 10, 12, 13, 14, 15, 17]. This kind of models make it possible to calculate terminal voltage and current of the modelled device and internal temperature of this device. It is assumed that in the modelled device the uniform temperature distribution is observed. Such models consist of the electrical model, describing the dependence between terminal voltages and currents of the considered element, and of the thermal model, describing the dependence of the internal temperature of the element on the power dissipated in it.

The worked out electrothermal model of a ferrite core is dedicated for SPICE software and it has a form of the subcircuit. The network representation of this model is shown in Fig.1.

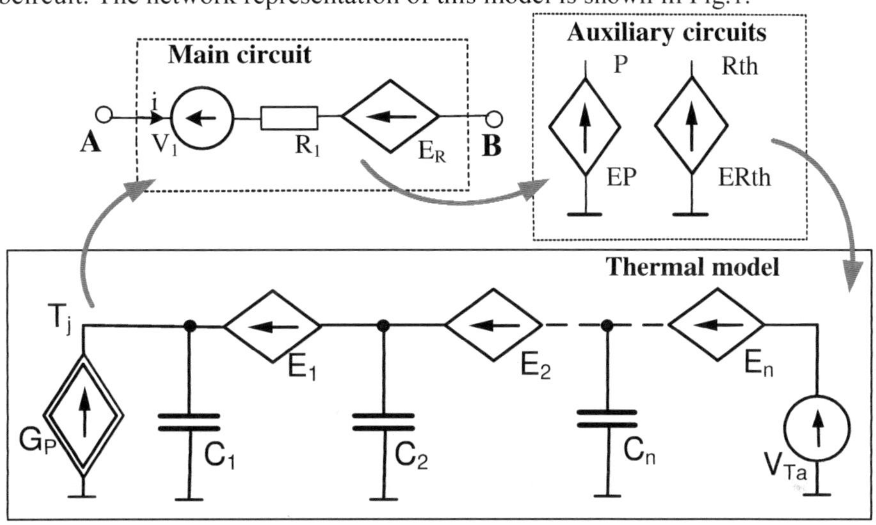


**Figure 1.** The network representation of the electrothermal model of a ferrite core

The presented model consists of two blocks: the electrical model and the thermal model. In the electrical model the voltage source $V_1$, the resistor $R_1$ and the controlled voltage source $E_R$ exist. This resistor represents the minimum value of resistance of the core. The voltage source $V_1$ of zero output voltage monitors the current of the core, whereas the voltage source $E_R$ describes the dependence of the core resistance on the current i flowing through the modelled core and on the core temperature $T_R$. The voltage on the voltage source $E_R$ is given by the following equation

$$E_R = \left[\left(m_1 \cdot \exp\left(-\frac{i}{n_1}\right) + m_2\right) \cdot \exp\left(B_1 \cdot \left(2 - \exp\left(-\frac{i}{n_2}\right)\right) \cdot \left(\frac{1}{T_R} - \frac{1}{T_0}\right)\right) + \left(m_3 \cdot \exp\left(-\frac{i}{n_3}\right) + m_4\right) \cdot \exp\left(B_2 \cdot \left(\frac{1}{T_R} - \frac{1}{T_0}\right)\right)\right] \cdot i \cdot R_0 \cdot \left(m_5 + m_6 \cdot \exp\left(-\frac{i}{n_4}\right)\right) \quad (1)$$

where $R_0$, $B_1$, $B_2$, $m_1$, $m_2$, $m_3$, $m_4$, $m_5$, $m_6$, $n_1$, $n_2$, $n_3$, $n_4$ and $T_0$ are parameters of the model.

In turn, in the thermal model the voltage on the terminal $T_R$ denotes the internal core temperature. This temperature is calculated with the use of the nonlinear thermal network of the form proposed in [18]. This network consists of linear capacitors representing thermal capacitances and controlled voltage sources representing thermal resistances of this core. The dependence of thermal resistance on the dissipated power is taken into account. The voltage source $V_{Ta}$ models the ambient temperature.

The voltage source $E_P$ represents the power dissipated in the core and its value is equal to the product of current and voltage of the core. The voltage source $E_{Rth}$ describes the dependence of the core thermal resistance on the dissipated power $p_R$. This dependence has a form given by [18]

$$R_{th} = R_{th0} + R_{th1} \cdot \exp\left(-\frac{p_R}{p_0}\right) \tag{2}$$

where $R_{th0}$, $R_{th1}$ and $p_0$ denote model parameters.

## 3. Manner of model parameters estimation

In order to use practically the elaborated model estimation of model parameters values is indispensable. These parameters are estimated using measurements and calculations.

Values of parameters describing electrical characteristics of the core are estimated by approximation of the measured dependences of the core resistance on current at the selected values of the ambient temperature $T_a$.

In turn, parameters existing in the thermal model are estimated using the procedure described in [8]. The measured waveforms of transient thermal impedance of the core are measured using the method proposed in the paper [19]. On the basis of the estimated values of the core thermal resistance at different values of the power dissipated in the core the values of parameters existing in Eq. (2) are estimated.

As a result of the estimation procedure the following values of model parameters of the core E22/6/16R made of material 3F3 are obtained: $R_0$ = 3.6 kΩ, $m_1$ = 0.53, $m_2$ = 0.38, $m_3$ = 0.61, $m_4$ = 0.17, $m_5$ = 0.65, $m_6$ = 0.7, $n_1$ = 3.2 mA, $n_2$ = 0.15 A, $n_3$ = 4.2 mA, $n_4$ = 0.125 A, $B_1$ = 1000 K, $B_2$ = 8500 K, $T_0$ = 300 K, $R_{th0}$ = 15.1 K/W, $R_{th1}$ = 11 K/W, $p_0$ = 5 W.

## 4. Results of calculations and measurements

The proposed model was verified experimentally for the planar ferrite core E22/6/16R made of material 3F3 by Ferroxcube [20]. This core is shown in Fig. 2.

**Figure 2.** The view of the investigated core with connectors

The results of calculations and measurements are presented in Figs. 3-8. In these figures the results of calculations are marked with lines, whereas the results of measurements - with points. During measurements the investigated core is situated in the thermal chamber with regulated temperature. In Figs. 3- 6 the dc electrical characteristics of this core are presented, whereas in Figs. 7-8 – transient characteristics are shown. During measurements the core is treated as an resistor and classical measurement set-up for the measure dc characteristics of this element is used. The temperature of the core is measured using pyrometer.

Fig.3 illustrates the dependence of resistance of the investigated core on the current at the selected values of the ambient temperature.

As it is visible, resistance of the core is a decreasing function of current and temperature. It is worth to notice that a change in temperature by about 130°C causes even a six-time change in the core resistance, and a change in the current value from 1 to 10 mA causes even a triple fall in this resistance. The observed changes in the value of the core resistance are a result only of changes in the ambient temperature, because the power dissipated in the core is near zero and self-heating phenomena are omittable.

Fig. 4 illustrates the influence of the ambient temperature on the current-voltage characteristics of

IOP Conf. Series: Journal of Physics: Conf. Series **1033** (2018) 012005 doi:10.1088/1742-6596/1033/1/012005

the considered core (Fig.4a) and the influence of the current on the dependence of resistance of the core on the ambient temperature (Fig.4b).

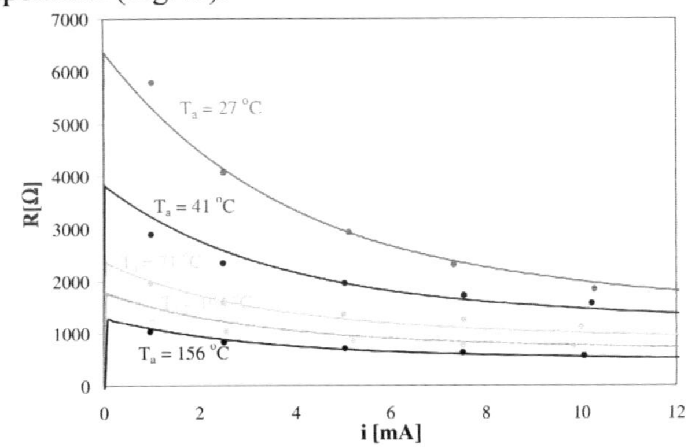


**Figure 3.** The measured and calculated dependences of the core resistance on the current at the selected values of the ambient temperature

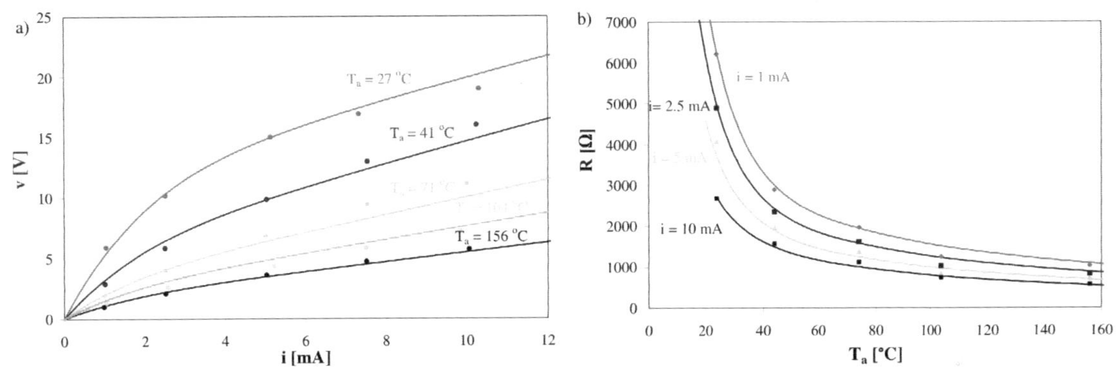


**Figure 4.** The measured and calculated current-voltage characteristics of the core (a) and the dependence of the core resistance on the ambient temperature (b)

As it is visible, both the ambient temperature and the current strongly influence on the resistance of the core. In the considered range of changes in these quantities the value of this resistance changes even six times.

In Fig. 5 the calculated and measured current-voltage characteristic of the considered core and the dependence of the core temperature on the voltage are shown.

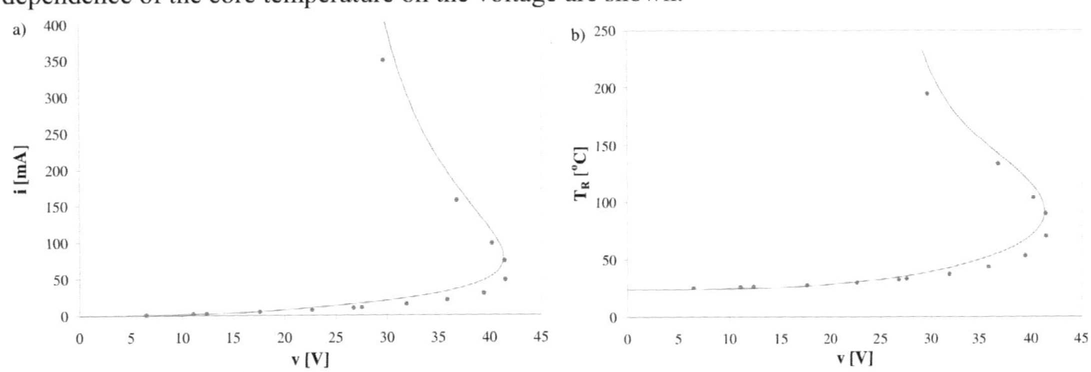


**Figure 5.** The measured and calculated current-voltage characteristic of the investigated core (a) and the dependence of the core temperature on the voltage (b)

As it is visible, the obtained dependence i(v) is not a function in the mathematical sense. In this characteristic the electrothermal breakdown is visible at the voltage exceeding 40 V. At this point the core temperature is equal to about 70°C. The temperature of the investigated core is an increasing function of the voltage and its value reaches even 200°C. The observed shape of the i(v) characteristics is a result of monotonically decreasing dependence of the core resistance on the current and a strong influence of temperature on this characteristic. The considered resistance decreases from about 7 kΩ at the current i = 1 mA to only 60 Ω at i = 0.35 A. An increase of the core temperature is slower and slower when the power dissipated in the core increases. This is a result of a decreasing dependence of the core thermal resistance on the dissipated power, described in the model using Eq.(2). This dependence is shown in Fig.6.

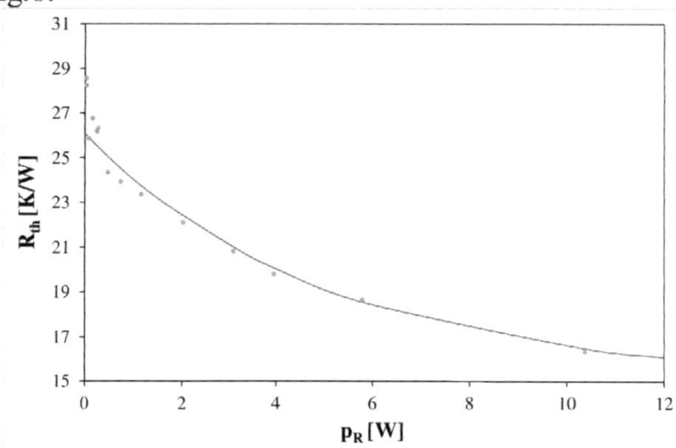


**Figure 6.** The measured (points) and calculated (line) dependence of the core thermal resistance on the dissipated power

The shape of this dependence results from the fact that efficiency of heat convection is a decreasing function of the difference between temperatures of the heat source and the surroundings [21].

Fig. 7 illustrates waveforms of the core transient thermal impedance obtained at the selected values of the dissipated power in the core. Changes equal even to 30% of the value of considered waveforms at the steady state is observed. The cooling conditions during the measurements of dc and the transient characteristics changes. Therefore, in the transient analysis the values of parameters $R_{th0}$ and $R_{th1}$ are equal to 16.1 K/W and 12 K/W, respectively. The steady state is observed after 3000 s from switching-on the power. The process of core heating is visible for time t > 100 s.

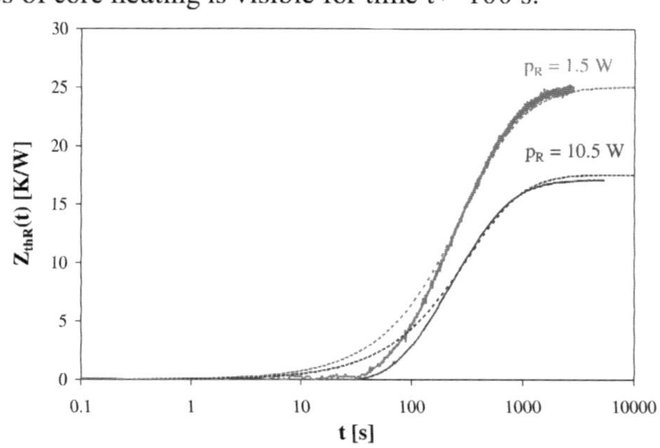


**Figure 7.** The measured (solid lines) and calculated (dashed lines) waveforms of transient thermal impedance of the core at the selected values of dissipated power

Fig.8 presents the measured waveforms of the core temperature (Fig. 8a) and the core resistance (Fig.8b). These waveforms were measured while the core was included in the circuit consisting of the voltage source $V_{SUP}$ connected in series with the resistor $R_S$ and the investigated core. This voltage source generates the signal of the shape of the voltage step of the value equal to $V_{SUP}$. Investigations were performed for 3 values of $V_{SUP}$ equal to 40 V, 50 V and 60 V, respectively.

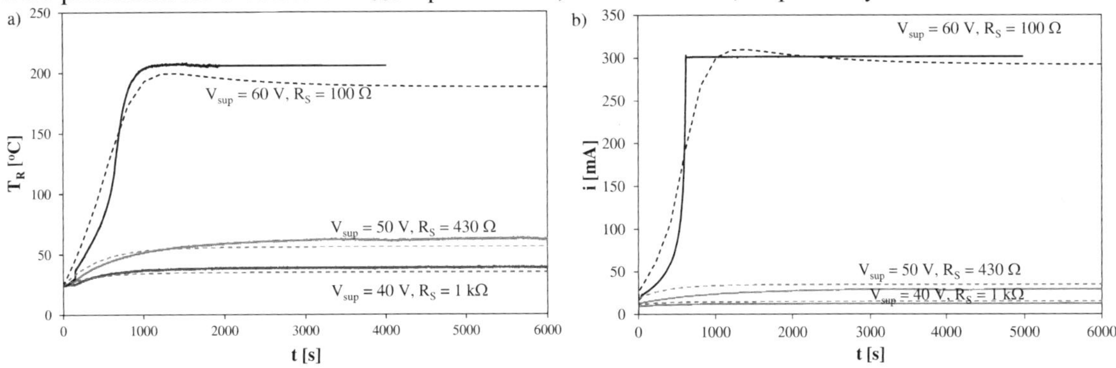


**Figure 8.** The measured (solid lines) and calculated (dashed lines) waveforms of the core temperature (a) and the core current (b) at different values of $V_{SUP}$

The observed waveforms of the core temperature are increasing functions, whereas the waveforms of the core resistance are decreasing functions. On the temperature waveforms the steady state is observed about 2000 s after the voltage jump. Maximum observed value of the core temperature exceeds even 200 °C. In turn, the core resistance decreases from over 4 kΩ to about 100 Ω only. The time indispensable to obtain the steady state of the dependence i(t) is the longest for $V_{SUP}$ = 50 V and the shortest for $V_{SUP}$ = 60 V.

## 5. Conclusions

In the paper the form of the electrothermal model of the ferrite core is proposed, in which both electrical phenomena and self-heating effect are taken into account. It was shown that this model assured a good agreement between the results of calculations and measurements both at dc characteristics and transient waveforms.

In the presented characteristics of the ferrite core it is visible that the dependence of resistance of the considered core on the current has the shape similar to a NTC thermistor. Additionally, it is visible that in the dc characteristics of this core the electrothermal breakdown occurs. It is proper also to notice that as a result of the flow of the direct current through the considered core, its temperature increases even to 200°C, and an increase of this temperature causes a significant decrease in resistance of this core. This can have an essential influence on the value of eddy currents induced in the core when this element operates in inductors or transformers. When the core temperature increases, eddy current should be taken into account.

The presented model and the results of investigations could be useful for designers of magnetic devices with ferrite cores. It will be possible to correctly calculate power losses in the core using the proposed model of the ferrite core.

## 6. References

[1] Van den Bossche A, Valchev VC 2005 Inductors and transformers for power electronics *CRC Press, Taylor & Francis Group* (Boca Raton)

[2] Ericson R, Maksimovic D 2001 Fundamentals of Power Electronics *Kluwer Academic Publisher* (Norwell)

[3] Wilson PR, Ross JN, Brown AD 2002 Simulation of magnetic component models in electric circuits including dynamic thermal effects *IEEE Transactions on Power Electronics* **17** (1) 55-65

[4] Wilson PR, Ross JN, Brown AD 2001 Optimizing the Jiles-Atherton model of hysteresis by a genetic algorithm *IEEE Transactions on Magnetics* **37** (2) 989-993

[5] Jiles DC, Atherton DL 1986 Theory of ferromagnetic hysteresis *Journal of Magnetism and Magnetic Materials* **61** 48-60.

[6] Bertotti G, Fiorillo F, Soardo G 1988 The prediction of power losses in soft magnetic materials. *Journal de Physique*, Colloque C8, Tome **49**

[7] Barlik R, Nowak M, Grzejszczak P, Zdanowski M 2016 Estimation of power losses in a high-frequency planar transformer using a thermal camera *Archives of Electrical Engineering* **65** (3) 613-627

[8] Górecki K, Rogalska M, Zarębski J 2014 Parameter estimation of the electrothermal model of the ferromagnetic core *Microelectronics Reliability* **54** (5) 978-984

[9] Górecki K, Górecki P 2017 Modelling dynamic characteristics of the IGBT with thermal phenomena taken into account *Microelectronics International* **34** (3) 160-164

[10] Mawby PA, Igic PM, Towers MS 2001 Physically based compact device models for circuit modelling applications *Microelectronics Journal* **32** 433 – 447

[11] Singh J 2001 Semiconductor Devices. Basic Principles. *John Wiley & Sons, Inc.*(New York)

[12] Górecki K 2015 Modelling mutual thermal interactions between power LEDs in SPICE *Microelectronics Reliability* **55** (2) 389-395

[13] Starzak Ł, Zubert M, Janicki M, Torzewicz T, Napieralska M, Jabłoński G, Napieralski A 2013 Behavioral approach to SiC MPS diode electrothermal model generation *IEEE Transactions on Electron Devices* **60** (2) 630-638

[14] Górecki K, Detka K 2012 Electrothermal model of choking-coils for the analysis of dc-dc converters *Materials Science & Engineering B* **177** (15) 1248-1253

[15] Górecki K, Zarębski J 2009 Electrothermal analysis of the self-excited push-pull dc-dc converter *Microelectronics Reliability* **49** (4) 424-430

[16] Godlewska M, Górecki K, Górski K 2016 Modelling the pulse transformer in SPICE *IOP Conference Series: Materials Science and Engineering* **104** 39th International Microelectronics and Packaging IMAPS Poland 2015 Conference 012011 1-11 doi:10.1088/1757-899X/104/1/012011

[17] Janke W 1992 Zjawiska termiczne w elementach i układach półprzewodnikowych *WNT* (Warszawa)

[18] Górecki P, Górecki K 2017 Non-linear compact thermal model of IGBTs *41st International Microelectronics and Packaging IMAPS Poland 2017 Conference*, Warszawa, Full Contributtion_i114_157

[19] Górecki K, Górski K 2017 Compact thermal model of planar transformers *Proceedings of the 24th International Conference Mixed Design of Integrated Circuits and Systems Mixdes 2017* Bydgoszcz 345-350

[20] 3F3 Material specification 2008 *Ferroxcube*

[21] Yener Y, Kakac S 2008 Heat Conduction *Taylor &Francis*

# Researches on Tie Rod Ends Lubricated by Grease with $TiO_2$ and $ZrO_2$ Nanoparticles

**Marek Wozniak[1], Krzysztof Siczek[1] *, Przemysław Kubiak[1], Piotr Jozwiak[1], Krystian Siczek[2]**

[1]Lodz University of Technology / Department of Vehicles and Fundamentals of Machine Design, Stefanowskiego Str. 1/15, 90-924 Lodz, Poland,
[2]Lodz University of Technology / Faculty of Electrical, Electronic, Computer and Control Engineering, Stefanowskiego Str. 18/22, 90-924 Lodz, Poland

*E-mail: krzysztof.siczek@p.lodz.pl

**Abstract**. The nanoparticles of some materials can be used successfully to improve tribological properties through decreasing both wear and friction borne out of contact between the contact surfaces of elements in different devices, particularly vehicles. Nanoparticles of $TiO_2$ and $ZrO_2$ were chosen as additives to the lithium grease lubricating the contact surfaces in tie rod ends. The object of study was the steel ball – the component of the tie rod end – mating with the polymer insert and lubricated with the pure lithium grease or containing the addition of pure $TiO_2$, pure $ZrO_2$ nanoparticles, with a 1%wt. Studies on friction were carried out using the tester allowing cyclical rotational motion and different loading of contact. Wear was investigated by driving a car, whose tie rod ends were analysed, on a fixed 'eight'-shape track and with a fixed velocity pattern. The aim of the study was to obtain the values and waveforms of friction moment and wear versus cycles, loading and composition of lubricating grease. The waveforms of friction coefficient were obtained using the FEM model of the analysed contact zone. Based on the obtained waveforms, recommendations for the composition of additives for lithium grease were made.

## 1. Introduction

Recently, interest has increased regarding the use of nanoparticles as a lubricant additive for various devices, as in [1]. Nanoparticle additives can reduce friction and wear through various mechanisms. According to ref. [2] such mechanisms are classified as direct and indirect effects of nanoparticle additives on the friction and wear process. Studies described in ref. [3–5] show that nanoparticle additives play the role of ball bearings between friction surfaces, resulting in decreased friction and wear. Further studies, described in [6–8], show that nanoparticle additives can also form a protective film over friction surfaces. Study [9] describes the formation of nanoparticle additive depositions on contact surfaces which compensate the loss of material. Nanoparticle additives can also reduce the roughness of contact surfaces due to their abrasive action as described in [10]. Study [11] presents research on the effect of nanoparticle additives on journal bearings using the finite element method. Study [12] presents investigations on the effects of nanoparticle lubricant additives on the performance of externally adjustable fluid film bearings. The studies described in [11, 12] point to an increase in the load-carrying capacity and stability of journal bearings using a nanoparticle-dispersed lubricant, and use the viscosity data reported in [5] for $TiO_2$, CuO and nano-diamond dispersions in base oil to compute the bearing characteristics. The variation in lubricant viscosity with changes in nanoparticle concentrations and the effect of nanoparticle size and its aggregation properties were omitted during these studies. Paper [13] describes a novel method for evaluating the load-carrying capacity of journal bearings operating on lubricants containing nanoparticle additives. It was found that variations in lubricant viscosity due to the nanoparticle additives can be simulated accurately using a modified Krieger-Dougherty viscosity model. Study [13] presents research on the effect of the viscosity variation of engine oil, following the addition of $TiO_2$ nanoparticles at volume fractions ranging from 0.005 to 0.025, on the load-carrying capacity of a journal bearing. The $TiO_2$ nanoparticles, even at low concentrations of 0.01 volume fraction, could increase the load-carrying capacity of journal bearings by 40% in comparison to plain engine oil without nanoparticle additives. The $TiO_2$ nanoparticles of primary size <~100 nm dispersed in engine oil formed aggregates of an average size of 777 nm, forming a particle packing fraction of 7.77. The load-carrying capacity of the journal bearing operating on a $TiO_2$-based nanolubricant at a constant volume fraction increased with higher nanoparticle aggregate packing ratios. Increasing the particle packing fraction from 7.77 to 10 resulted in a 35% increase in the load-carrying capacity for a $TiO_2$ nanoparticle concentration of 0.015 volume fraction.

The studies on effect of $TiO_2$ as an additive to oil are presented in [14, 15].

The studies on the effect of TiO2 nanoparticles as an additive to liquid paraffin are presented in [16].

In studies reported in [17], $TiO_2$/CuO nanoparticles were dispersed into lithium grease to improve its lubricating properties. Then, the process of sliding friction was tested. The friction force of the contact interface between a ball and a flat surface with $TiO_2$/CuO nano-grease was measured for a ball on a flat sliding wear tester. The tested lithium grease was synthesised with 0.5wt.%, 1.0wt.%, 1.5wt.% and 2.0wt.% $TiO_2$/CuO nanoparticles. It was found that the $TiO_2$/CuO nanoparticle additives can improve the wear resistance of lithium grease and decrease the friction coefficient. A 1.0 wt.% content of $TiO_2$ nanoparticles was the best in reducing friction and wear, and reduced the friction coefficient of the rubbing interface by about 40%, according to the friction tests. The wear of the lithium grease with a 2.0wt.% content of added CuO nanoparticles, reduced wear by about 60% in comparison with pure lithium grease. Furthermore, the addition of components creating tribo-film and $TiO_2$/CuO nanoparticles on the worn surface also decreased friction and wear.

Study [18] presents investigations on the friction and wear of lithium grease using the $ZrO_2$ nanoparticles as the additive, aimed at studying the change of the tribological properties of lithium grease. The friction coefficient decreased by over 30% when $ZrO_2$ nanoparticles with mass fractions of 0.25% and 0.5% were added. The $ZrO_2$ nanoparticles with mass fraction of 0.25% and 0.5% achieved the best results, and the wear scar diameter decreased by 48.8% and 46.5% respectively.

In almost all vehicles, one can find the ball joint of the tie rod end (Figure 1a, as in [19]), which is a moveable connection (kinematic pair) enabling a rotationally oscillating movement of the connected components relative to one another. The axis of rotation of a wheel coincides with the axis of the pin. This enables angular deflection and transmits the shearing and longitudinal forces (along the pin axis). Ball joints perform a rotationally oscillating movement, and for this they are lubricated with solid lubricant through a grease nipple, or with graphite grease when a grease nipple is absent.

The kinematic scheme of the steering mechanism is presented in Figure 1b. The steering angle $\delta_f$ is the independent variable. Force $F$ loading the steering rod is the function of geometrical parameters: $H_C$ – length of the tie rod, $\alpha_C$ – auxiliary angle, $h_C$ - length of the steering knuckle arm, $R_C$ – radius approximately equal to the distance of the middle point of contact between the wheel and the road surface from the axis of the steering knuckle, lateral offset at the ground $s$, lateral inclination angle $\lambda$, caster angle $\nu_C$, tire radius $r$, vehicle mass $m$, rolling resistance coefficient $f_t$, air density $\rho$, drag coefficient $C_x$, reference frontal area $A$, vertical load of wheel $F_z$, vehicle speed $v$.

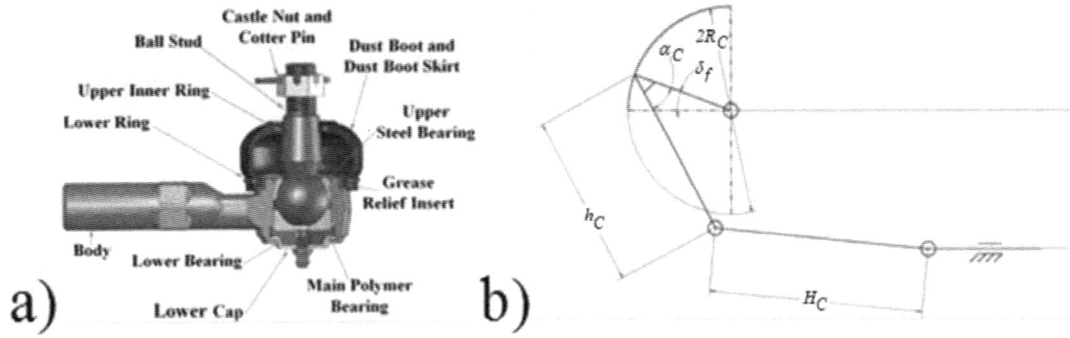


**Figure 1.** a) Components of the tie rod end, as given in [19], b) The kinematic scheme of the steering mechanism.

The aim of the study was to obtain the values and waveforms of friction moment and wear versus cycles, and the loading and composition of lubricating grease for the contact between the steel ball and polyethylene bearing of the tie rod end.

## 2. Materials and methods

### *2.1. Samples preparation*

Samples for researching the friction coefficient include the rotating ball mating with the fixed counterpart. A ball with a diameter of 18 mm was made out of steel and polished. It rotated with constant speed and was loaded by the tribotester head with a constant force. The mating counterpart contained the polyethylene sample immersed in an epoxy resin cast and chemically cured in the steel ring. The counterpart inner diameter was 30 mm. The contact zone was lubricated with lithium grease with or without the addition of 1% wt. of $ZrO_2$ or $TiO_2$ nanoparticles. The $ZrO_2$ nanoparticles were manufactured by the Institute of High Pressure Physics of the Polish Academy of Sciences. Their size was 10–20 nm, as reported in [20]. The $TiO_2$ nanoparticles were introduced via the commercial product NanoTitanium delivered by Lodz Amepox, reported in [21]. Such a product was formulated with nanosize $TiO_2$ particles of 10 nm in size, with a concentration of 1%wt. dispersed in isopropanol, which was evaporated directly before the application of $TiO_2$ into the grease. The grease was mixed with the commercial product using the High-Speed Planetary Mixer THINKY SR-5000, reported in [22], and the amount resulted in a concentration of 1% wt. of nanoparticles in the lithium grease. This was then treated with ultrasonic waves in the GT Sonic VGT-800 Ultrasonic Cleaner, for only about 2 min., to avoid agglomeration or the structural break of the nanoparticles.

### *2.2. Researching the friction coefficient*

The friction coefficient in the contact zone between the steel ball – the component of the tie rod end mating with the polymer insert and lubricated with the lithium grease containing different nanoparticles – could only be estimated, due to the available tribotester, described in [23] (Figure 2). This tester allowed the measurement of the spinning friction torque in the contact zone between ball 4 and its counterpart 5, based on the twisting angle of string 7 supported by the aerostatic bearing of table 9. The twisting angle was read on shield 8. Measurement was done at a constant rotating speed of shaft 1 equal to 36 rpm and constant loading from the set of weights 3 increased in a step manner for the consecutive series 7.2 to 27.2 N. The abovementioned estimation of the friction coefficient was made using the modelled contact zone between steel ball 4 and its counterpart 5 made out of the same material as the original bearing. The force loading of the ball and rotating speed were chosen so that similar averaged values of contact pressure and sliding speed as in the original case would be obtained. Using the mentioned tribotester, reported in [23], the waveforms of friction torque against time were obtained for the 1%wt. amount of $TiO_2$ and $ZrO_2$ nanoparticles in the lithium grease. The friction coefficient was calculated using measured values of the friction torque, diameter of contact zone and averaged contact pressure in the contact zone. The averaged contact pressure and diameter of the contact zone were estimated using the FEM model of such a zone (Figures 3a, 3b). The contact model had the assumed material properties of the steel ball and polymer counterpart, and the initial geometrical parameters and loading force of the real tribotester.

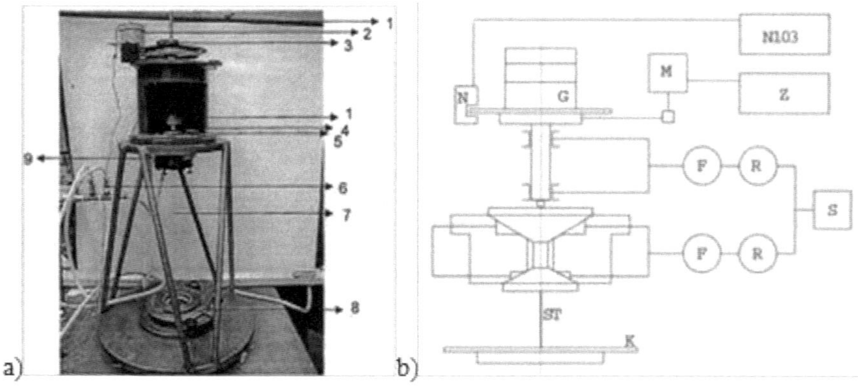


**Figure 2.** The ball-on-inner-sphere tribotester for measurement of the spinning friction torque; a) photograph of the tester. 1 – shaft, 2 – electric motor, 3 – weights, 4 – ball, 5 – inner sphere or plane counterpart, 6 – compressed air supply, 7 – string, 8 – shield for measurement of twisting angle, 9 – table with aerostatic bearing, as in [23]; b) Scheme of the tester. F – filter, G – weight, M – electric motor, N – tachometer, K – shield for measurement of twisting angle, S – compressor, ST – string, R – reducer, Z – motor power supply.

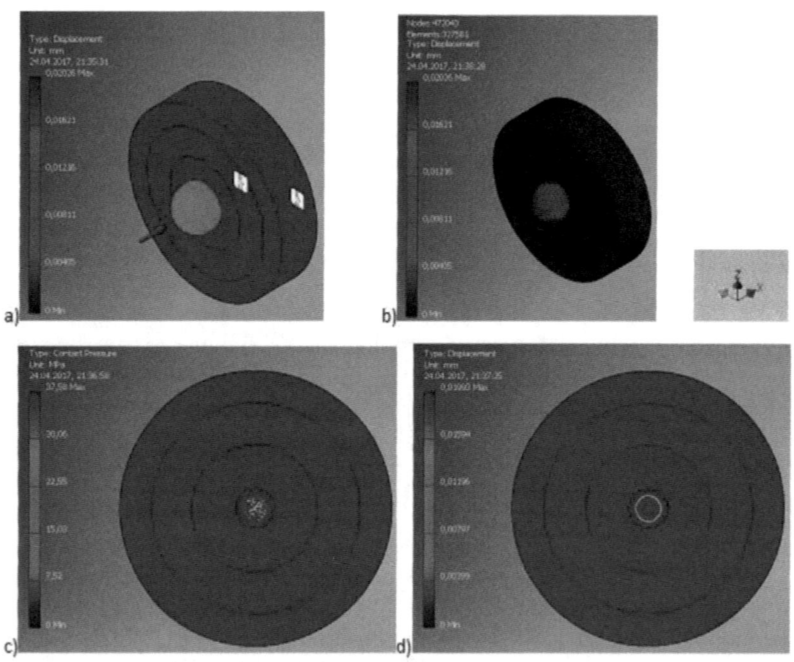


**Figure 3.** a) Model of steel ball – polymer insert contact, b) Network of finite elements, c) Contact pressure, d) Displacements.

### *2.3. Researches of wear coefficient*

It was assumed, for simplicity, that the linear wear intensity $I_h$ for both the steel ball stud and the polymer bearing is of the form (1):

$$I_h = kp_a^2 = \begin{cases} k_1 p_a^2, \; for\;steel\;ball\;stud \\ k_2 p_a^2, \; for\;polymer\;bearing \end{cases} \tag{1}$$

where:
$k_1$ and $k_2$ - the proportional factor for the steel ball stud and polymer bearing, respectively,
$p_a = F/S$ – average contact pressure,
$F$ – force loading contact zone,
$S$ – contact area between contact surfaces.

The wear intensity factor $k$ characterizing the linear wear intensity was estimated from the equation (2):

$$\begin{aligned} W = w_z \cdot t_u = I_h \cdot S \cdot \textstyle\sum_{j=1}^{j=N}[v_{aver}(j) \cdot t_u(j)] = (k \cdot p_a^2) \cdot S \cdot \sum_{j=1}^{j=N}[v_{aver}(j) \cdot t_u(j)] = \\ = k \cdot F \cdot p_a \cdot \textstyle\sum_{j=1}^{j=N}[v_{aver}(j) \cdot t_u(j)] \rightarrow k = \frac{W}{F \cdot p_a \cdot \sum_{j=1}^{j=N}[v_{aver}(j) \cdot t_u(j)]} \end{aligned} \tag{2}$$

where:
$w_z$ – volumetric wear rate,
$j$ – consecutive number of drive cycle in series,
$N$ – number of cycles in series.

The wear intensity was calculated using values of the measured volumetric wear, and calculated: the averaged sliding speed and the averaged contact pressure. The volumetric wear was investigated while driving a car with tie rod ends analyzed, on the track of the fixed 'eight'-shape and with the fixed velocity pattern (Figure 4). Each series contains $N$ = 1000 cycles. The bearing was lubricated with analysed grease, applied into contact zone via a grease nipple specially mounted to the tie rod (Figure 5). The total wear value could be obtained from the measured diameters: the ball outer diameter $D_1$ and the bearing inner diameter $D_2$ before and after test series. As the changes of steel ball diameter $D_1$ were negligible, the volumetric wear resulted only from the changes of the bearing inner diameter.

The wear $W_{seat}$ of polymer bearing was calculated as a difference between volumes contained within the worn and new bearing and bordered by its inner surface. It was assumed, that the shape of new bearing volume is a sphere of diameter $D_2$ = 22 mm. The worn seat volume is a rotated ellipse with two main diameters equal $D_2$ and the third main diameter equal the sum $D_2 + \Delta D_2$. The $\Delta D_2$ is the change of the measured inner diameter due wear process occurring between the ball end of the steel stud and polymer seat. So, the wear of ball end of the steel stud is estimated from the equation (3):

$$W_{seat} \approx \frac{4}{3 \cdot 8} \cdot \pi \cdot D_2^2 \cdot \Delta D_2 \tag{3}$$

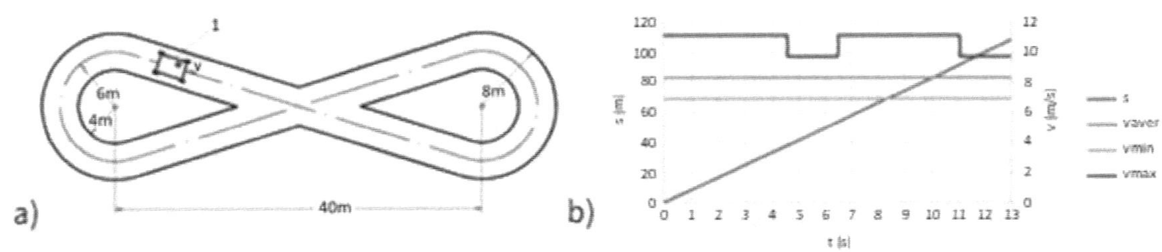


**Figure 4.** a) The track of the fixed 'eight'-shape, b) waveforms of vehicle velocity v.

**Figure 5.** Modified tie rod end with grease nipple.

The contact pressure was obtained from the steady state analysis using the FEM model of contact between rotating steel ball stud and its polymer bearing fixed in the steel insert (Figure 4). The network of finite elements was generated automatically by the program. The boundary conditions were following:
$C_1$ - the cylindrical surface of the part the steel insert was fixed.
$C_2$ - the top surface of the ball stud could move only perpendicularly to the axis the stud.
$C_3$ - the conical surface of the ball stood was loaded by the constant horizontal force $F$.
The network of FE and boundary conditions were presented in the Figure 6.

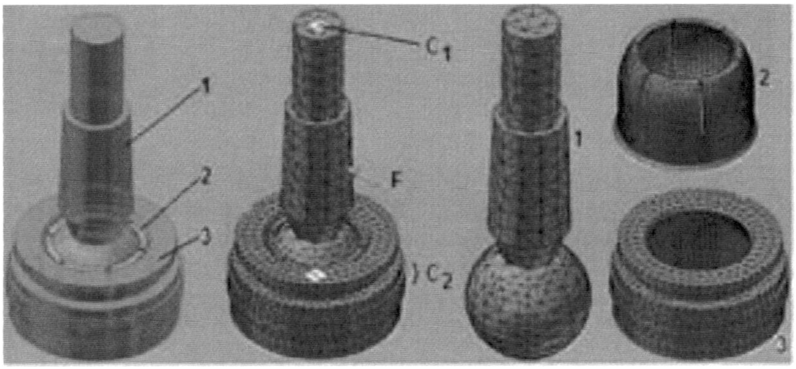


**Figure 6.** FEM model of contact between rotating steel ball stud 1 and its polyethylene bearing 2 fixed in the steel insert 3.

Averaged pressure was calculated from the equation (4):

$$p_a = \frac{1}{M}\sum_{e=1}^{e=M} p_e \quad (4)$$

where:
$p_e$ – contact pressure on the $e$-th contact element, $M$ – number of contact elements loaded by contact pressure (based on status number: 1 – in contact, 0 – not in contact, assigned to each contact element after solution convergence).

The horizontal force $F$ loading the ball stud was estimated using measurement by dynamometer positioned as in the Figure 7.

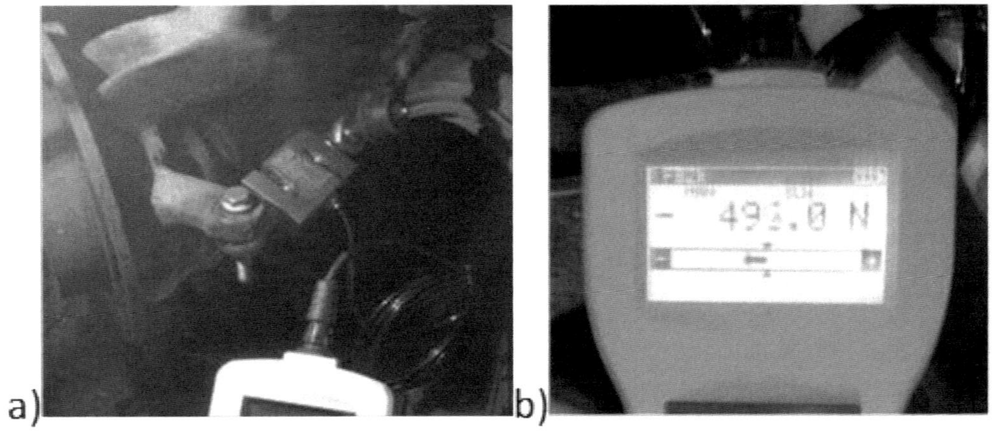


**Figure 7.** The measurement of horizontal force F loading the ball stud. a) spring sensor, b) measuring unit.

*2.4. The scheme of cornering*

To estimate the average sliding speed of ball stud against the polyethylene bearing, the scheme of cornering was assumed. During first one fifth of the turn length $l_{max}$ the driver turned the steered wheels to the needed value of angle, namely $\alpha_{max}$. The next three-fifths of the turn length he kept the steered wheels with such needed value. During last one fifth of the turn length the driver turned the steered wheels back to the initial positions. Such algorithm was illustrated in the Figure 8.

The time $t_{corn}$ of cornering can be estimated from the equation (5):

$$t_{corn} \approx \frac{l_{max}}{v_{aver}} = \alpha_{max} \cdot \frac{R}{v_{aver}} \tag{5}$$

The time of turning the steered wheel forward and back is equal $0.4 \cdot t_{corn}$.

The average sliding speed $v_b$ of ball stud relative to the polyethylene bearing is estimated from the equation (6):

$$v_b = \frac{2 \cdot \alpha_{max} \cdot D_1}{0.4 \cdot t_{corn}} \tag{6}$$

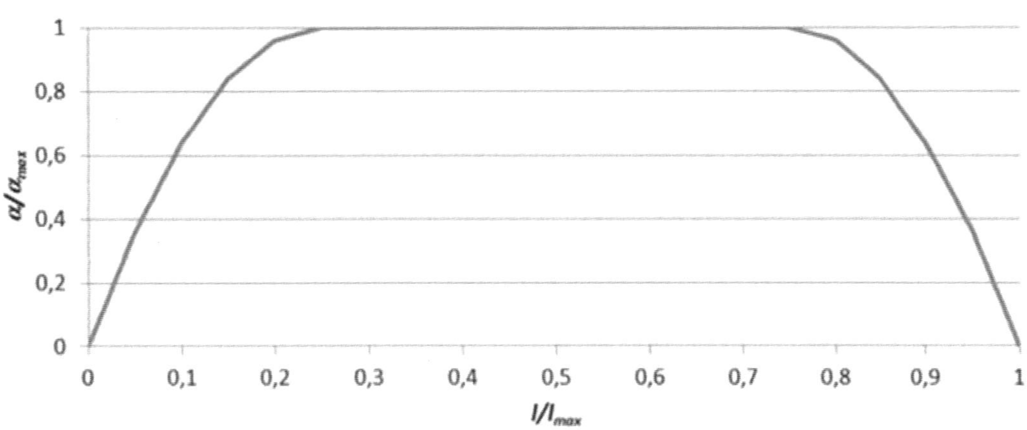


**Figure 8.** The scheme of cornering.

IOP Conf. Series: Journal of Physics: Conf. Series **1033** (2018) 012006 doi:10.1088/1742-6596/1033/1/012006

## 3. Results and discussion

The contact pressure distribution in the polymer bearing obtained from the FEM model of contact between the ball stood and its bearing was presented in the Figure 9. Average value was of 1.6 MPa and maximal about 4.6 MPa.

The values of the volume wear W for the polymer bearing and calculated corresponding values of parameter k is equal 48 mm$^3$ for $ZrO_2$ and 32 mm$^3$ for $TiO_2$ were shown in the Table 1. It was observed, that the addition of 1%wt both the $ZrO_2$ and $TiO_2$ into the pure lithium grease can decrease twice the volume wear in contact between the steel ball of stud and its polyethylene bearing in the test conditions. It was in agreement with the report given in [17] for the $TiO_2$ nanoparticles and also with the report given in [18] for the $ZrO_2$ nanoparticles, although the amount of $ZrO_2$ in the lithium grease was twice lower, than for the present study.

**Figure 9.** Contact pressure distribution in the polyethylene bearing.

**Table 1.** The values of the volume wear *W* for the polymer bearing and calculated corresponding values of parameter *k* for cases of pure lithium grease and lithium grease with addition of $ZrO_2$ and $TiO_2$ nanoparticles.

| Grease | Volume wear W [mm$^3$] | Parameter *k* [1/MPa$^2$] |
|---|---|---|
| Lithium | 1.143 | 0.285 |
| Lithium + 1%wt $TiO_2$ | 1.143 | 0.285 |
| Lithium + 1%wt $ZrO_2$ | 2.571 | 0.109 |

The friction coefficient against force F loading contact zone between the steel ball mating with the polyethylene inner sphere is presented in the Figure 10. The values of the friction coefficient in contact between steel ball and polyethylene inner sphere nonlinearly decreased with the increase of the force F loading contact zone for investigated lubrication cases: with pure lithium grease, lithium grease with the addition of 1%wt $ZrO_2$ nanoparticles and lithium grease with the addition of 1%wt $TiO_2$ nanoparticles. The decrease is the highest for addition of 1%wt of $ZrO_2$. It was also in agreement with results reported in [17] for the TiO2 nanoparticles and also with those reported in [18] for the $ZrO_2$ nanoparticles, although the amount of $ZrO_2$ in the lithium grease was twice lower, than for the present study.

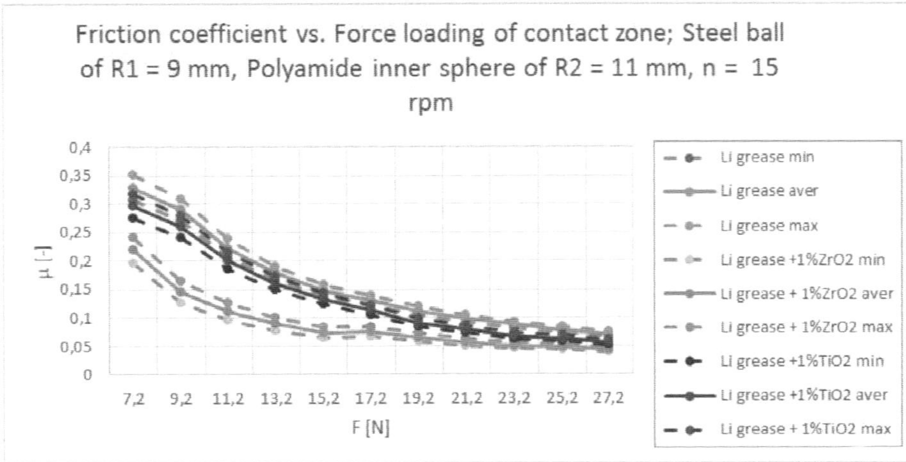


**Figure 10.** The friction coefficient against force loading contact zone between the steel ball mating with the polyethylene inner sphere for the case of lubrication by: the pure lithium grease, the lithium grease with additive of 1%wt of $ZrO_2$ nanoparticles and the lithium grease with additive of 1%wt $TiO_2$ nanoparticles.

## 4. Conclusions

1. For all investigated cases of lubrication for components the tie rod end, the coefficient of friction in contact between ball surface of steel stud and inner spherical surface of polyethylene bearing nonlinearly decreased with the increase of the force *F* loading contact zone.
2. The addition of 1%wt $ZrO_2$ nanoparticles into pure lithium grease decreased the mentioned coefficient of friction by up to 50%, especially for lower values the loading force *F*.
3. The addition of 1%wt $TiO_2$ nanoparticles into the pure lithium grease decreased the mentioned coefficient of friction by up to 20 % in the almost whole range of the loading force *F* tested.
4. The addition of 1%wt both the $ZrO_2$ and $TiO_2$ into the pure lithium grease decreased twice the volumetric wear occurring in contact between the steel ball of stud and its polyethylene bearing in the test conditions.

## References

[1] Zhang Z J, Simionesie D and Schaschke C 2014 Graphite and Hybrid Nanomaterials as Lubricant Additives *Lubricants* **2** 44-65
[2] Lee K, Hwang Y, Cheong S, Choi Y, Kwon L, Lee J and Kim S H 2009 Understanding the role of nanoparticles in nano-oil lubrication *Tribology Letters* **35(2)** 127 – 131
[3] Rapoport L, Leshchinsky V, Lvovsky M, Nepomnyashchy O, Volovik Y and Tenne R 2002 Mechanism of friction of fullerenes *Industrial Lubrication and Tribology* **54(4)** 171-176
[4] Chinas-Castillo F and Spikes H A 2003 Mechanism of action of colloidal solid dispersions. Transactions of the American Society of Mechanical Engineers *Journal of Tribology* **125** 552 – 557
[5] Wu Y Y, Tsui W C and Liu T C 2007 Experimental analysis of tribological properties of lubricating oils with nanoparticle additives *Wear* **262** 819 – 825
[6] Hu Z S, Lai R, Lou F, Wang L G, Chen Z L, Chen G X and Dong J X 2002 Preparation and tribological properties of nanometer magnesium borate as lubricating oil additive *Wear* **252** 370 – 374
[7] Ginzburg B M, Shibaev L A, Kireenko O F, Shepelevskii A A, Baidakova M V and Sitnikova A A 2002 Antiwear effect of fullerene C60 additives to lubricating oils *Russian Journal of Applied Chemistry* **75(8)** 1330 – 1335
[8] Xiaodong Z, Xun F, Huaqiang S and Zhengshui H 2007 Lubricating properties of Cyanex 302-modified MoS2 microspheres in base oil 500SN *Lubrication Science* **19(1)** 71 – 79

[9] Liu G, Li X, Qin B, Xing D, Guo Y and Fan R 2004 Investigation of the mending effect and mechanism of copper nano-particles on a tribologically stressed surface *Tribology Letters* **17(4)** 961 – 966
[10] Tao X, Jiazheng Z and Kang X 1996 The ball-bearing effect of diamond nanoparticles as an oil additive *Journal of Physics D: Applied Physics* **29** 2932 – 2937
[11] Nair K P, Ahmed M S and Al-qahtani S T 2009 Static and dynamic analysis of hydrodynamic journal bearing operating under nano lubricants *International Journal of Nanoparticles* **2** 251 – 262
[12] Shenoy B S, Binu K G, Pai R, Rao D S and Pai R S 2012 Effect of nanoparticle additives on the performance of an externally adjustable fluid film bearing *Tribology International* **45(1)** 38 – 42
[13] Binu K G, Shenoy B S, Rao D S and Pai R 2014 A variable Viscosity Approach for the Evaluation of load Carrying Capacity of oil lubricated Journal Bearing with $TiO_2$ Nanoparticles as lubricant Additives *Procedia Materials Science* **6** 1051 – 1067
[14] Ilie F and Covaliu C 2016 Tribological Properties of the Lubricant Containing Titanium Dioxide Nanoparticles as an Additive *Lubricants* **4** 12
[15] Ilie F, Covaliu C and Chisiu G 2014 Tribological Study of ecological Lubricants Containing Titanium Dioxide Nanoparticles *Applied Mechanics and Materials* **658** 323-328
[16] Zhang L, Chen L, Wan H, Chen J and Zhou H 2011 Synthesis and tribological properties of stearic acid-modified anatase (TiO2) nanoparticles *Tribological Letters* **41** 409–416
[17] Chang H, Lan C W, Chen C H, Kao M J and Guo J B 2014 Anti-Wear and Friction Properties of Nanoparticles as Additives in the Lithium Grease *International Journal of Precision Engineering and Manufacturing* **15(10)** 2059-2063
[18] Xia X T, Zhang Y P, Zhang Y Z and Chen S C 2010 Influence of $ZrO_2$ Nanoparticle as Additive on Tribological Property of Lithium Grease *Applied Mechanics and Materials* **26-28** 83-87
[19] http://www.deezachassis.com/parts_tre.html Available at: 2017.07.07
[20] Nanoparticles Zirconium Oxide ($ZrO_2$) https://www.unipress.waw.pl Available at: 2017.07.07
[21] NanoTitanium. General description http://amepox.com.pl/en/products-with-nanotitanium/ Available at: 2017.07.07
[22] Planetary Centrifugal Mixer "THINKY MIXER" AR-100 http://www.thinkyusa.com/products/item-all/rotation-revolution-mixer/ar-100.html Avalable at. 2017.07.07
[23] Burcan J 2010 The Questions of Spinning Friction in an Elastic Point of Contact *Tribologia* **4** 47-56

# The influence of PVD/CVD processes parameters on Ni catalyst nanoparticles sizes

**Izabela Stępińska[1], Elżbieta Czerwosz[1], Mirosław Kozłowski[1] and Piotr Dłużewski[2]**

[1]Tele and Radio Research Institute, Ratuszowa 11, 03-450 Warsaw, Poland [2]Institute of Physics PASc., al. Lotników 32/46, 02-668 Warsaw, Poland
E-mail: izabela.stepinska@itr.org.pl

**Abstract**. In this work we demonstrate how technological parameters of PVD/CVD method influence the growth of Ni particles' sizes where carbon nanotubes films are grown. In CVD process we changed the growth temperature while other parameters were kept strictly constant. Carbon nanotubes (CNTs) were grown on carbonaceous-nickel substrates by chemical vapor deposition (CVD) of xylene gas at the temperature range 600–700°C. Ni particles' sizes affect diameters, length and distribution density of carbon nanotubes obtained in different temperatures of CVD processes.

## 1. Introduction

Nickel is a metal which has been used successfully as a catalyst in chemical vapor deposition process for the growth of thin films of various materials. In this paper we present the role of Ni metal particles in the formation of nanotubes synthesized by CVD (Chemical Vapor Deposition) method. For synthesizing CNTs nanometer-size metal particles are required to enable hydrocarbon decomposition at a lower temperature than the spontaneous decomposition temperature of the hydrocarbon. Fe, Co and Ni are most commonly used metals for two main reasons: (1) high solubility of carbon in these metals at high temperatures; and (2) high carbon diffusion rate in these metals [1]. The unusual ability of the transition metals to produce graphitic carbon is considered to be related to a number of factors which include their catalytic ability to decompose the volatile carbon source, the formation of metastable carbides and the diffusion of carbon through metal catalyst particles [2]. Growth mechanism of CNTs is connected to the diffusion of carbon atoms through nanometer-scale catalytic particles. These carbon atoms form growing nanotubes [3-6]. Growing mechanism (tip- growth, bottom- or root-growth as well as open tip-growth) depends on a type of catalyst, process temperature, time and flowing hydrocarbon kind. Additionally, depending on the size of the catalyst particle, a graphite grains, carbon filaments, or carbon nanotubes can form. As more carbon is deposited on the catalyst, it will either diffuse into or over the surface of the particle and become incorporated into the graphitic lattice, increasing the tube's length. When a catalyst particle remains fixed to the substrate, the process is called root growth. When it is detached from the surface and remains encapsulated within the opposite end, it is called tip growth. There are several theories on the growth mechanism for nanotubes [7-9]. Some of them suggest that metal catalyst particles are floating or are supported on graphite substrate to form or to be incorporated into growing nanotube. The ability of nickel to catalyze CNTs formation is mostly linked to it catalytic activity for the decomposition of carbon compounds, its ability to form carbides and the possibility for carbon to diffuse through and over the metals extremely rapidly [10-13]. In CVD, metal catalyst nanoparticles are prepared on substrate surface which is put in the furnace.

The diameter of the CNTs formed usually depends on the physical dimensions of the metal catalyst. In [14] authors suggest that the diameter of CNTs can be controlled by the size of the catalyst particle. They described a method in which actually the outer diameter of the nanotubes matched the original particle size. It was the HFCVD method with a low growth temperature of 550°C, where the particles did not coalesce on the substrate during the CNT growth. Ref. [15] presents carbon:nickel nanocomposite templates (NCTs) which were used as catalyst precursors for diameter-controlled growth of single-walled carbon nanotubes (SWCNTs) by chemical vapor deposition . SWCNT diameter distribution and the presence of multi-walled CNTs were analyzed by SEM and Raman spectroscopy. It is demonstrated that SWCNT diameters strongly depend on the NCT type applied for CVD synthesis.

## 2. Experimental

### *2.1. Synthesis of carbon nanotubes*

The nickel nanoparticles used as a catalyst for the growth of carbon nanotubes were prepared on a Si substrate by Physical Vapour Deposition (PVD) method [16]. PVD process was performed from two separated sources containing fullerene $C_{60}$ (99.95%) and nickel acetate under a dynamic vacuum of $10^{-3}$ Pa. All these initial films were obtained in the same conditions i.e. deposition time t = 8 min, current intensity through both sources $I_{C60} = 1.9$ A, $I_{Ni} = 1.1$ A, and sources-substrate distance d = 69 mm. These particles with diameter of a few nanometers, were embedded on carbonaceous matrix.

The CVD process was carried out under dynamic conditions in an argon and xylene flow (40 L/h) in a quartz tube. In a heating zone of an oven, at a high temperature, decomposition of the hydrocarbon and chemical reaction on the substrate follows. Ni in the form of small nanoparticles, obtained during the PVD process, at high temperature agglomerates into larger grains which migrates to the surface of the film and stimulates the growth of nanotubes.

In CVD process the temperature was changed at constant duration time and xylene doze. The time was 30 mins and xylene doze was 0.1ml/min wherein temperature was 600, 650 and 700°C.

### *2.2. Characterization methods*

The temperature influence was investigated by SEM and TEM microscopy. SEM studies with the use of backscattered and secondary electrons for these films were performed with JEOL JSM-7600F field emission scanning electron microscope, operating at 5 keV incident energy, equipped with EDX system. The topography of the film surface was imaged with a secondary electrons (SE) detector. A low-angle backscattered electron (LABE) detector enabled to define the morphology of the samples by observing changes in the contrast composition. The contrast depends on the atomic number of the elements of the observed microstructure. The areas with a higher atomic number are brighter, while those with a lower atomic number are darker. The analysis of the Ni content was performed by X-ray spectrometry.

TEM studies and electron diffraction were performed with a JEOL JEM-2000EX electron transmission microscope operating at 200 keV electron beam incident energy.

## 3. Results and discussion

### *3.1. Ni nanoparticles after PVD and CVD process*

Nanocomposite films obtained by PVD process contain a lot of minor nickel nanograins (Fig. 1a). The histogram of the particles sizes (Fig. 1b) demonstrates that most of them have diameters between 4-6 nm. Nanoparticles are distributed inside in the carbonaceous matrix that is built of amorphous carbon grains and fullerite grains. The average thickness of nanocomposite films obtained by PVD process is 250 nm.

To identify the structure of films, HR TEM (High Resolution TEM) imaging was used, which made it possible to measure the interplanar distances of the phases present there in. Fig. 2a and 2b depict images of high resolution crystallographic planes that allow to measure them (enlarged areas of fields marked with a yellow frame). The measured interplanar distances on the test portion of the film correspond to the array value for nickel with the crystalline type *fcc* for the planes with the indices (111)

- Fig. 2a. On the other hand, the interplanar distance measured on the carbon matrix fragment, which is shown in Fig. 2b, corresponds to the spacing between planes (111) for the fullerite.

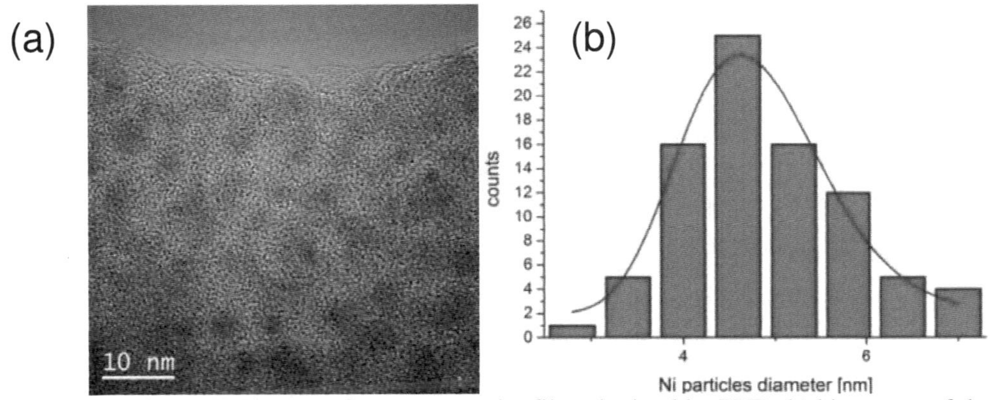


**Figure 1** a) TEM image of nanocomposite film obtained by PVD, b) histogram of the Ni nanoparticles size.

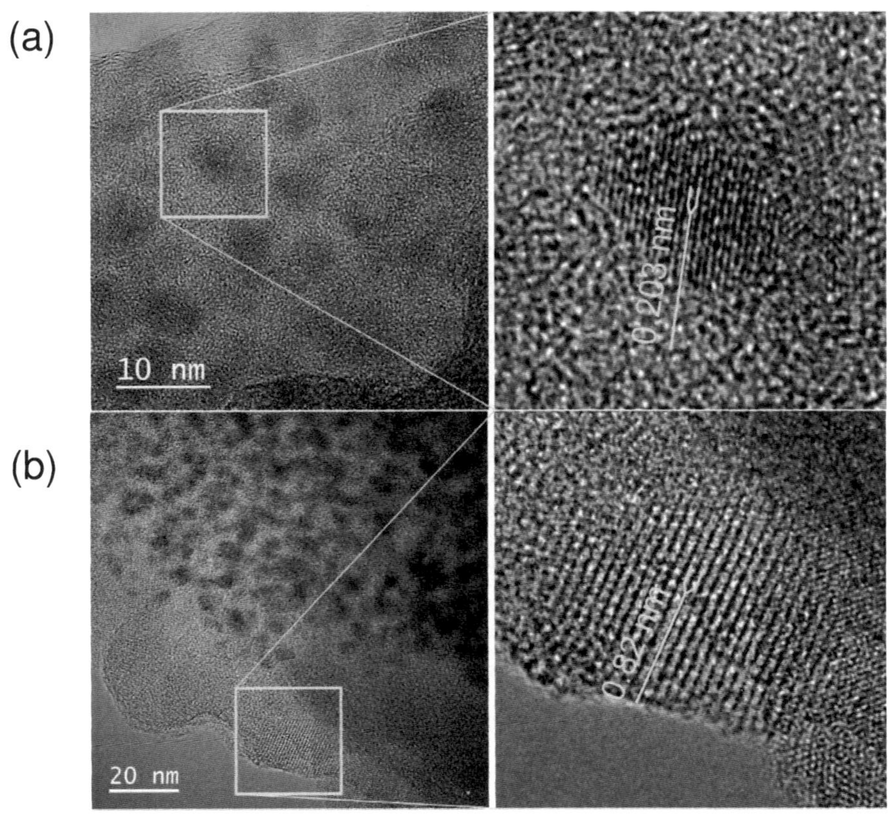


**Figure 2** TEM image of a) Ni nanoparticles, b) area of carbonaceous matrix

Fig. 3 presents topography of carbonaceous-nickel film. The topography demonstrates that the surface is composed of C-Ni grains which are huge (up to 200 nm) and have rectangular shape. Table 1 presents the Ni content for films obtained by PVD process. Beyond the film component additional peak originating from the substrate (Si) appeared. The presence of oxygen is due to the oxidation of the film surface.

IOP Conf. Series: Journal of Physics: Conf. Series **1033** (2018) 012007 doi:10.1088/1742-6596/1033/1/012007

**Table 1** Ni content from EDS measurements

| C | O | Si | Ni | Total |
|---|---|---|---|---|
| 80.46 | 2.90 | 1.20 | 15.43 | 100.00 |

*All results in weight%

**Figure 3** SEM image of the topography of nanocomposite film obtained by PVD

The nanocomposite films were used as initial substrates for CNTs growth. In the modification by CVD process these particles diffuse to the films' surface under the influence of high temperature. During such a migration nickel coalesces into bigger grains and becomes a catalyst of CNTs growth. After CVD process Ni grains have diameters between 20-60 nm and they depend on technological parameters of CVD process. SEM and TEM images analysis show that the tip growth of carbon nanotubes takes place in CVD process. Comparison of SEM images in SE and LABE mode leads up to the conclusion that nickel nanograins (bright dots in Fig. 4b) exist inside carbonaceous matrix (dark background in Fig. 4b) and in the core of carbon nanotubes, especially on the top of them. While a nanotube grows Ni grains can divide and migrate within a forming nanotube. It was observed that carbon can diffuse into Ni grains and we suppose that this effect could cause division of metal grain. During CVD process due to high temperature and interaction between Ni and carbon atoms, Nickel grain divides into smaller fragments which are retained in various places of nanotube (Fig. 4). HR TEM images of CNT fragment also prove that obtained by PVD/CVD method CNTs have multiwall structure (Fig. 4c).

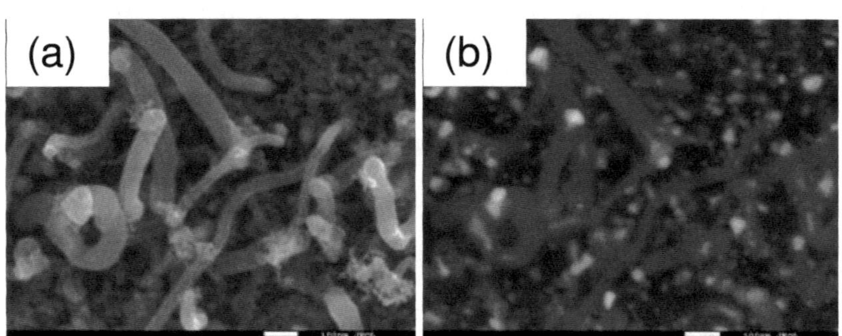

IOP Conf. Series: Journal of Physics: Conf. Series **1033** (2018) 012007 doi:10.1088/1742-6596/1033/1/012007

(c)

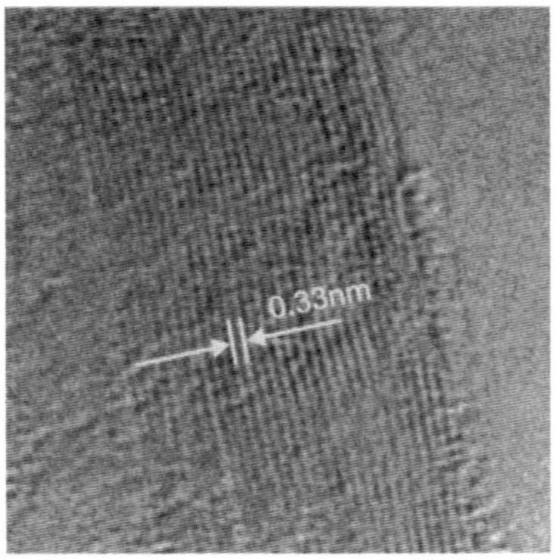


**Figure 4** SEM image of the morphology and structure of CNTs film obtained by PVD/CVD in a) SE mode and b) LABE mode; c) HR TEM image of CNT fragment with visible graphite plane

*3.2. The influence of CVD process on Ni particles diameters*

In CVD process the temperature increased while other parameters were kept strictly constant. Figures 5, 6, 7 present carbon nanotubes films obtained in three different CVD processes where temperature was 600, 650 and 700°C, duration time 30 mins and xylene doze 0.1 ml/min. SEM images of CNTs films obtained by backscattered electrons demonstrate Ni nanoparticles as bright dots (due to their higher atomic number) and carbon objects as dark background or shapes. The histograms of Ni particles diameters were prepared on the base of SEM images obtained in LABE mode. Special computer program for analysis of microscopic images was applied for this purpose.

*3.2.1. CVD process at 600°C.* Carbon nanotubes are not observed in a film obtained at 600°C. There are only Ni particles distributed very closely to the film's surface. From the histogram of particles diameters (Fig. 5b) it was estimated that the average size of Ni catalysts is 22.5 nm. From TEM images it is visible that particles have rounded shapes (Fig. 6a) and are surrounded by carbon shells (Fig. 6b). The interplanar distance measured on a fragment of Ni nanoparticle is 0.33 nm, which is shown in Fig. 6b, and it corresponds to the spacing between planes of graphite. When the carbon shells form on the catalyst particles, the nucleus of growth process of CNTs is blocked. The carbon shell encapsulates the Ni particle, what stops the activity of catalyst preventing nanotube's growth. There are also some objects with elongated shapes present and they have Ni particles inside too. These objects are probably carbon nanotubes at the very early step of growth that ended before the carbon clusters had grown up.

(a)

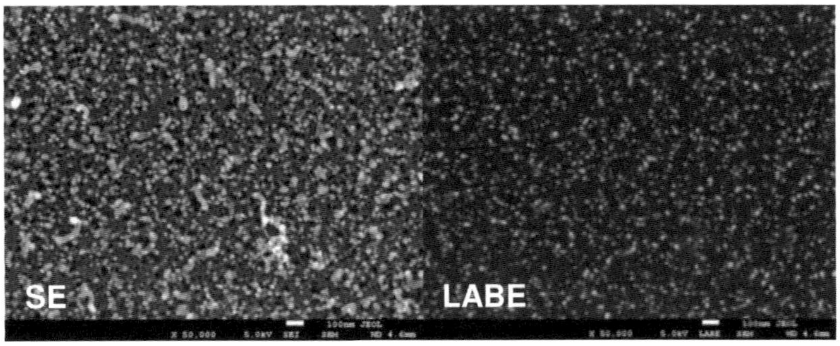

(b)

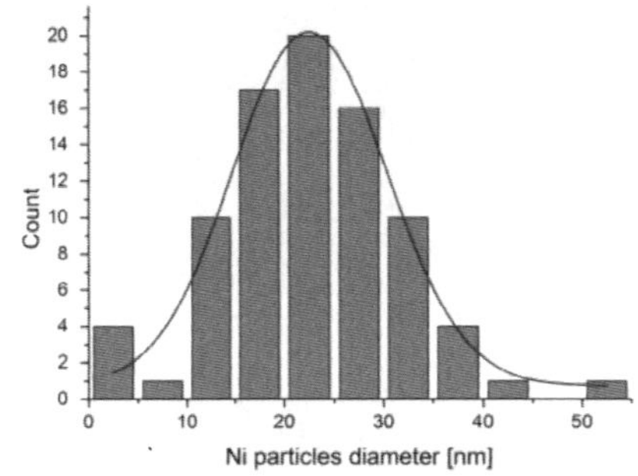


**Figure 5** a) SEM image of the topography of CNTs films obtained by PVD/CVD at 600°C, b) histogram of Ni particles diameter

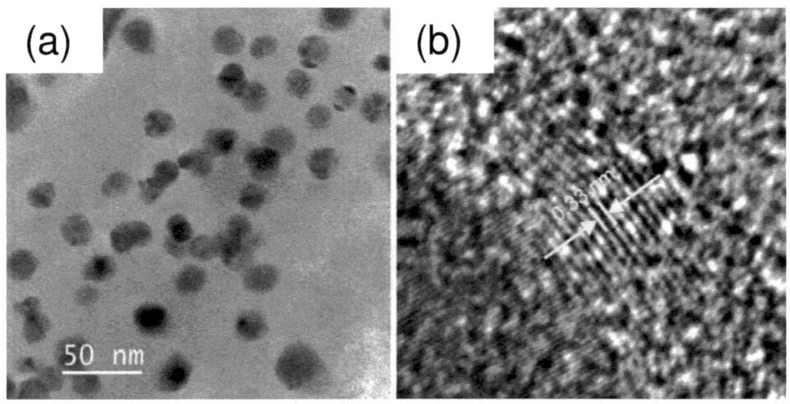


**Figure 6** a) TEM image of carbon nanotubes film obtained at 600°C, b) a fragment of Ni nanoparticle surrounded by graphite like shells

*3.2.2. CVD process at 650°C.* Fig. 7 shows that film obtained at 650°C consists of many carbon nanotubes. CNTs are long and twisted. The diameters are various and have between 50 up to 100 nm. In case of the film obtained at 650°C from SEM image in LABE mode (Fig. 7a), it was found that Ni particles are placed in CNTs, especially on the top but also along of them. It is clearly visible in TEM images in high resolution (Fig. 8). Almost all nanotubes are ended by Ni particles with the average size 32 nm. At the same time much smaller nanoparticles are visible inside nanotubes. Various defects (nanotube bending, start of a new nanotube growth, discontinuity in graphite planes) appear in places where these smaller Ni particles are found. Many Ni grains are seen inside a nanotube and they are correlated to a formed defect in CNT form (Fig. 8b). From SEM images in LABE mode a histogram presenting size distribution of Ni nanograins was obtained. Most of Ni particles have diameters 30-40 nm, but also there are grains with bigger ones (50-60 nm). It seems that the higher temperature of CVD process causes faster migration to the film's surface and bigger coalescence of Ni nanoparticles.

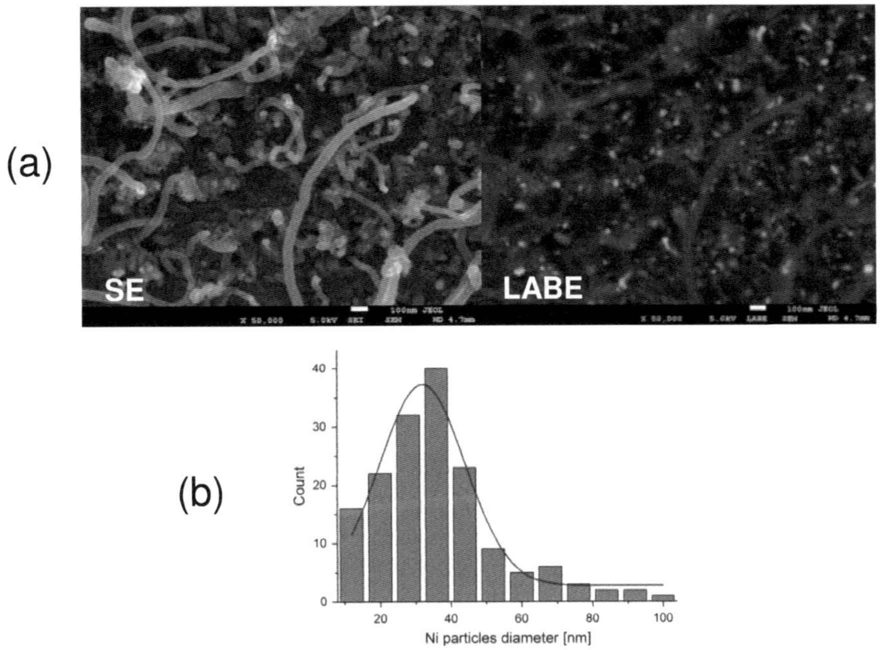


**Figure 7** a) SEM image of the topography of CNTs films obtained by PVD/CVD at 650°C, b) histogram of Ni particles diameter

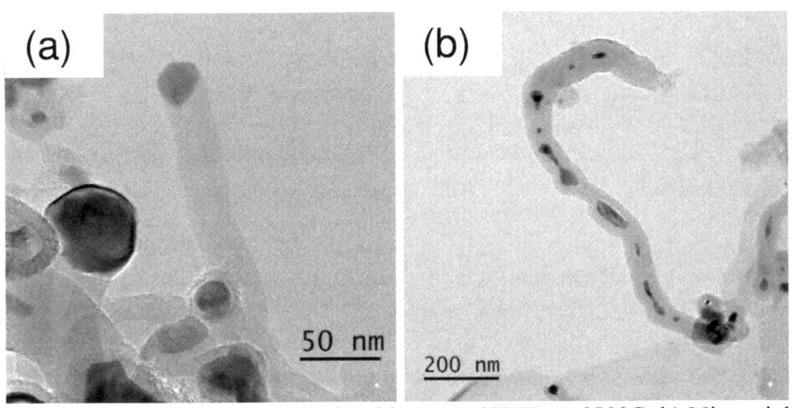


**Figure 8** a) TEM image of CNTs films obtained by PVD/CVD at 650°C, b) Ni particles inside CNT

*3.2.3. CVD process at 700°C.* Film obtained at the highest CVD process temperature is characterized by presence of rarely distributed CNTs on the surface (Fig.9a). From the histogram of particle diameters (Fig. 9b) it was estimated that Ni catalyst particles are much bigger then particles observed in films obtained be-low 700°C. The average size of Ni catalysts is 42 nm. The high temperature of CVD process causes coalescence of Ni nanoparticles into large grains. Smaller particles are also found and it is shown at the histogram of particles sizes. On the bigger particles the carbon shells are formed which block CNT growth like in the case of films obtained at 600°C. But visible particles with diameters below 40 nm can catalyze CNTs growth. Such particles occur in a small amount and CNTs film obtained at 700°C contains single nanotubes on the surface.

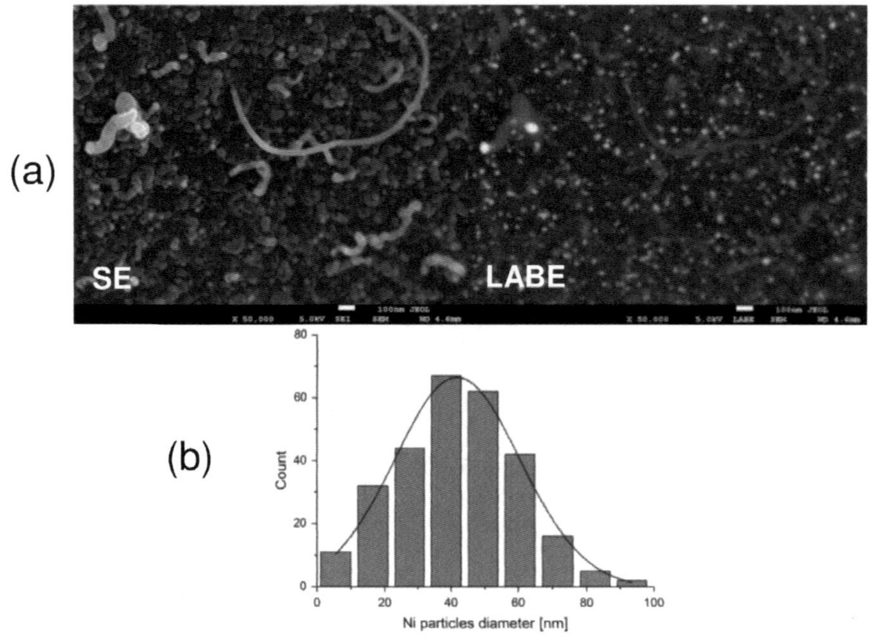


**Figure 9** a) SEM image of the topography of CNTs films obtained by PVD/CVD at 700°C, b) histogram of Ni particles diameter

## 4. Conclusions

It is widely known that nanotubes grow by the decomposition of the hydrocarbons gas on the surface of a catalyst particle. The carbon dissolves in the catalyst, diffuses through it, and exits to form the nanotube. Such an effect can only take place when catalyst particles have appropriate size. The catalyst particles that are too small cannot form a stable cap for nucleation and particles that are too big also do not nucleate. In such cases the carbon shells are formed and the growth of CNTs cannot occur since they are simply encapsulated by the precipitating carbon.

Our studies showed that technological parameters of PVD/CVD method had significantly influenced Ni catalyst nanoparticles. Nickel nanoparticles found in films obtained by PVD process have diameters 4-6 nm with crystalline form of *fcc* type. CVD process causes agglomeration of small particles into larger ones and their migration to the film' surface. It occurs for all samples obtained by CVD process. The process temperatures affect the intensity of this effect. The particles' size increases when the temperature grows. The nature of the catalyst and its size determine many properties of the CNTs obtained, such as the diameter of CNTs which grows when Ni nanoparticles sizes increase. Our investigation of carbon nanotube obtained in the PVD/CVD method also showed that these nanotubes were growing according to tip growth mechanism where Ni nanoparticle occurred at the ends of CNTs and divided into smaller particles. Such division retains the metal particle along nanotube and causes defects.

## Acknowledgements

This work was supported by the Applied Research Programme of the National Centre for Research and Development No PBS1/A5/16/2012 named "Light sources with cold emitter" and Ministry of Science and Higher Education under Research Project "Nanocomposite C-Ni-Pd films for opto- and bio-electronics applications.

## References

[1] Kunar M and Ando Y 2010 *J. Nanosci. Nanotechnol.* **10**, 3739

[2] Ducati C, Alexandrou I, Chhowalla M, Robertson J and Amaratunga G A J 2004 *J. Appl. Phys.* **95 (11)** 6387
[3] Kowalska E, Radomska J, Byszewski P, Lange H and Dłużewski P 2003 *J. Mater. Res.* **18 (10)** 2452
[4] Bronikowski M J 2006 *Carbon* **44** 2822
[5] Zhao N, He Ch, Jiang Z, Li J and Li Y 2006 *Materials Letters* **60** 159
[6] Musatov A L, Kiselev N A, Zakharow D N, Kukovitskii E F, Zhabanov A I, Izraelyants K R and Chirakova E G 2001 *Applied Surface Science* **183** 111
[7] Sinnott S B, Andrews R, Qian D, Rao A M, Mao Z, Dickey E C and Derbyshire F 1999 *Chem. Phys. Lett.* **315** 25
[8] Charlier J C and Iijima S 2001 *Growth mechanisms of carbon nanotubes Carbon Nanotubes Topics Appl. Phys.* **80** ed Dresselhaus M S, Dresselhaus G and Avouris P (Springer-Verlag, Berlin Heidelberg, 2001) pp. 55–81
[9] Zhang X, Cao A, Wei B, Li Y, Wei J, Xu C and Wu D 2002 *Chem. Phys. Lett.* **362** 285
[10] Rashid HU, Yu K, Umar M N, Anjum M N, Khan K, Ahmad N and Jan M T 2015 *Rev. Adv. Mater. Sci.* **40** 235
[11] Scheffler M, Greil P, Berger A, Pippel E and Woltersdorf J 2004 *Materials Chemistry and Physics* **84** 131
[12] Gore J P and Sane A 2011 *Flame synthesis of carbon nanotubes Carbon Nanotubes - Synthesis, Characterization, Applications* ed Yellampalli S (InTech, 2011) chapter 7 pp 121-146
[13] Khalilov U, Bogaerts A and Neyts E C 2015 *Nat. Commun.* **6** 10306
[14] Sato S, Kawabata A, Nihei M and Awano Y 2003 *Chemical Physics Letters* **382** 361
[15] Melkhanova S, Haluska M, Hübner R, Kunze T, Keller A, Abrasonis G, Gemminga S and Krause M 2016 *Nanoscale* **8** 14888
[16] Kowalska E, Czerwosz E, Dłużewski P, Kozłowski M and Radomska J 2004 *Diamond and Related Mater.* **13** 1008

IOP Conf. Series: Journal of Physics: Conf. Series **1033** (2018) 012008 doi:10.1088/1742-6596/1033/1/012008

# Carburizing treatment of low alloy steels: Effect of technological parameters

**Younes Benarioua**[1, 2, *]

[1] Mechanical Engineering Department, Faculty of Technology, University of M'sila Bordj Bou Arréridj Read, M'sila-28000, Algeria.
[2] Emerging Materials Research Unit, University of Sétif 1, Sétif-19000, Algeria

* Corresponding author e-mail: benarioua younes@yahoo.fr

**Abstract.** The surface areas of the parts subjected to mechanical loads influence to a great extent the resistance to wear and fatigue. In majority of cases, producing of a hard superficial layer on a tough substrate is conducive to an increased resistance to mechanical wear and fatigue. Cementation treatment of low alloy steels which bonds superficial martensitic layer of high hardness and lateral compressive to a core of lower hardness and greater toughness is an example of a good solution of the problem. The high hardness of the martensitic layer is due to an increased concentration of interstitial carbon atoms in the austenite before quenching. The lower hardness of the core after quenching is due to the presence of ferrite and pearlite components which appear if the cooling rate after austenitization becomes lower than the critical on. The objective of the present study was to obtain a cemented surface layer on low alloy steel by means of pack carburizing treatment. Different steel grades, austenitization temperatures as well as different soaking times were used as parameters of the pack carburizing treatment. During this treatment, carbon atoms from the pack powder diffuse toward the steels surface and form compounds of iron carbides. The effect of carburizing parameters on the transformation rate of low carbon surface layer of the low alloy steel to the cemented one was investigated by several analytical techniques.

## 1. Introduction

Carburizing treatment carried out at a temperature between 1143°K and 1203 °K is used to increase the surface hardness of steel parts. The depth of carburization is a function of carburising time and the carbon potential at the surface [1-5]. High carbon potential is suitable for short carburising times [6-9]. The soaking step of carburizing treatment of low carbon steels is usually followed by quenching and tempering at low temperature [10, 11]. Carburization is the oldest heat treatment used for improving the wear and fatigue resistance of steel parts. For achieving this objective, it is necessary to know the type of steel in order to choose appropriate carburization parameters [12].

The steel grade, the soaking time and the temperature of the carburization were chosen as three technological parameters in this study. During this treatment, carbon atoms from the powder pack diffuse toward the steel surface and further on to the subsurface zone where they combine with the iron atoms to form the iron carbides. Effect of carburizing time and temperature on the transformation rate of low alloy steel to a cemented steel layer was investigated by means of different experimental techniques. The growth kinetics of carburized layer by diffusion-precipitation as a function of the soaking temperature, duration of the carburization and the steel grade are presented in the work.

## 2. Experimental details

Two low alloy steels (AFNOR 18CN4 and 22MC4) were used for carburization. The chemical compositions of the investigated steel grades are given in Table 1.

**Table 1.** Chemical composition of the investigated steel grades (wt. %)

| Grade | C | Ni | Cr | Mn | Si | Cu | Mo |
|---|---|---|---|---|---|---|---|
| 18CN4 | 0.18 | 0.4 | 1.02 | 0.09 | 0.028 | 0.03 | 0.05 |
| 22MC4 | 0.22 | 0.08 | 0.32 | 0.89 | 0.05 | 0.02 | 0.02 |

Cylindrical samples 20 mm in diameter and 10 mm long were cut out of the steels. In order to reduce their roughness to Ra = 0.25 μm, the samples were mechanically polished with use of sandpaper. Both steel substrates were received a pack cementation treatment. A cement coke powder (as a source of carbon atoms) mixed with barium carbonate (as an activator) was used as carburizing medium. The specimens were carburized at different temperature (1143, 1173 and 1203°K) and different soaking time (2, 4 and 6 hours).

To provide optimal mechanical properties, all samples after carburization underwent a quenching in oil and tempering at temperature of 443°K for a length of time of 2 hours. After carburizing, transverse cross-sections were made of the samples and the section surface was polished. The samples were then etched in a solution containing 3% of nitric acid in ethyl alcohol. Olympus microscope type B O71 was used to investigate the microstructure and the thickness of the carburized zone of the samples. Three measure experiments were performed in order to obtain a representative thickness value. The phase composition of the cemented layer was investigated by X-ray diffraction using Phillips diffractometer. Cu Kα characteristic radiation with a wavelength of 0.154 nm was used over a 2Θ range of 25–95°. The hardness profile was measured the cross-sections along the direction normal to the surface of the cemented layer by a Leco microhardness tester equipped with a Vickers indenter. The hardness was obtained on specular surface of the samples by applying 0.25 N of indentation load.

## 3. Results and discussion

### *3.1. Structure*

X-ray diffraction spectra of 22MC4 steel carburized at temperature of 1203°K for time of 2, 4 and 6 hours are shown in Fig. 1. X-ray diffraction pattern corresponding to sample treated for 2 hours shows that only small volume fraction of iron carbides, $Fe_2C$ and $Fe_3C_2$ were formed in the cemented layer.

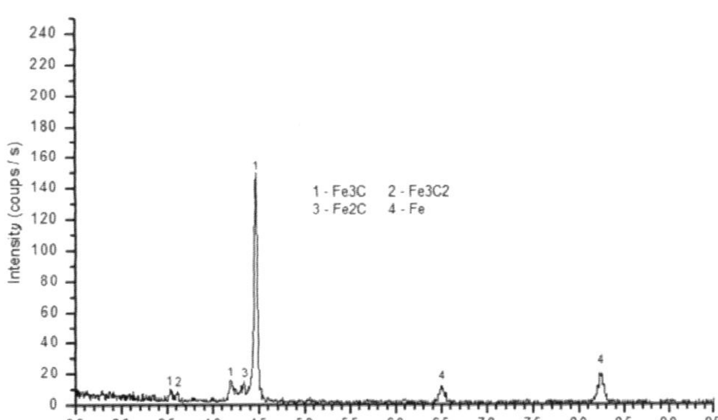


(A)

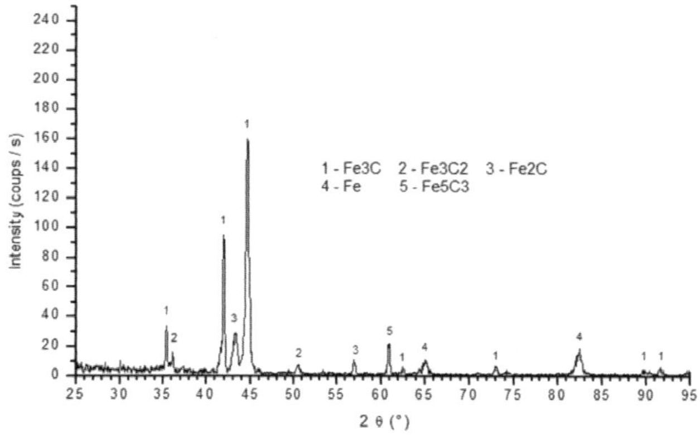


**(B)**

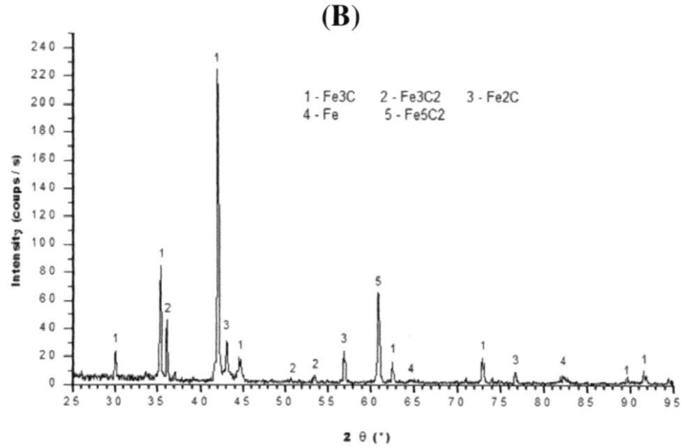


**(C)**

**Fig. 1.** X-Diffraction spectra of 22MC4 low alloy steel carburized at 1203°K for the length of time of 2 h (a), 4 h (b) and 6 h (c)

Two broad peaks from the tempered martensite phase of a low diffraction intensity appear in the diffraction spectra. The volume fraction of the iron carbides in the cemented layer increases progressively with carburizing temperature and the length of the soaking time. When the carburizing time increases, the volume fraction of the iron carbides increases as well. It follows from the X-ray diffraction spectra that a small amount of a new phase $Fe_5C_2$ appeared for carburizing time of 4 hours which however increased considerably for length of carburization time of 6 h. Contrary to this behaviour, it looks like that the volume fraction of the tempered martensite phase seems to decrease for the length of 6 hours of carburization.

*3.2. Morphology*

Micrographs of polished and etched sections of 18NC4 and 22MC4 steel carburized for the length of time of 4 hours at temperature of 1203 °K are shown in Figure 2. Three distinct zones were identified in the cross sections of two steel grades. An outer zone consists of a martensite rich in carbon and of different iron carbides, an intermediate region consists a diffusion zone of a lower content of carbon atoms and an inner region characteristic for the investigated steel grades not influenced by diffusion of carbon atoms from the carburizing medium. As indicated in table 2, the thickness of carburizing layers of steels increases with increasing of temperature and time of treatment

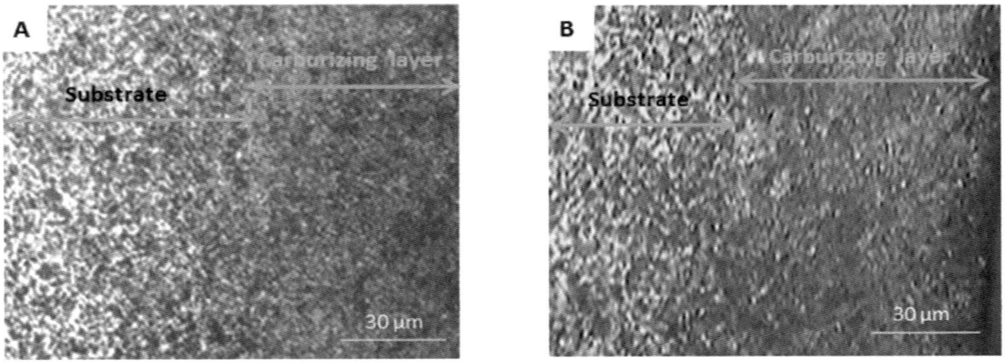


**Fig. 2.** Light micrographs of the structure of cemented layer on 18CN4 steel (A) and 22MC4 one (B) after carburizing treatment for 2 hours at 1173 °K

**Table 2.** Thickness of the cemented layers on 18CN4 steel and 22MC4 one for different temperature and the length of the carburizing time.

| T (°K) | | 18CN4 | | | 22MC4 | | |
|---|---|---|---|---|---|---|---|
| | Time (h) | 2 | 4 | 6 | 2 | 4 | 6 |
| 1143 | Thickness (µm) | 30 | 37 | 42 | 35 | 50 | 55 |
| 1173 | | 35 | 45 | 50 | 40 | 55 | 70 |
| 1203 | | 40 | 55 | 65 | 45 | 65 | 80 |

*3.3. Kinetic growth*

In the classical kinetic theory, the square of the cemented layer thickness as a function of the length of the carburizing time is described by [13-15].

$$d^2=Kt \tag{1}$$

where d is the thickness of the cemented layer, K is the kinetic growth rate constant and t the length of the carburizing time.

Figure 3 presents a relationship between the square of the thickness of the cemented layers produced on 16CN6 and 22MC4 steels and the length of the carburizing time for different temperatures. One can notice that the square of the cemented layer thickness for the both steel grades increases linearly with increasing length of carburizing time.

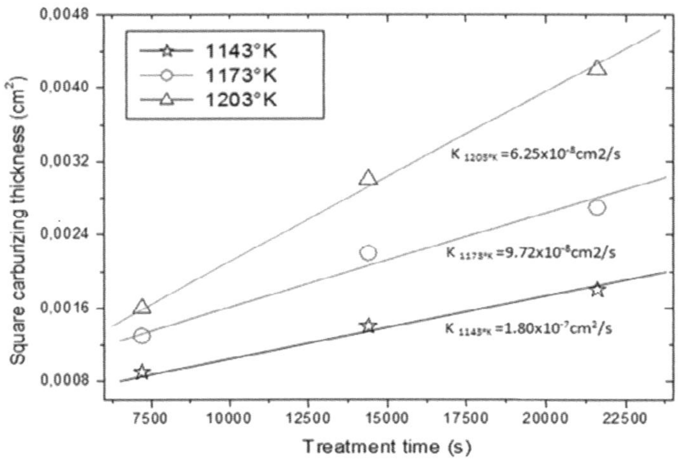


**(A)**

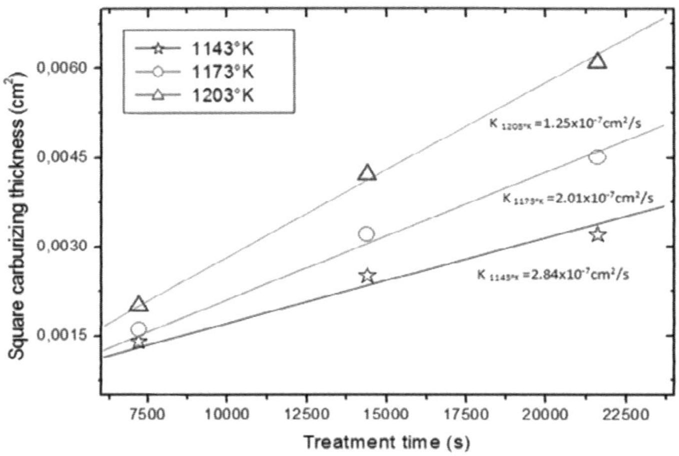


**(B)**

**Fig. 3.** Square of the cemented layer thickness as a function of the length of the carburizing time at different temperature of the carburizing process for 18CN4 steel (A) and 22MC4 one (B)

From the slopes of the plots indicated in Fig. 3, the growth rate constants $K_T$ of the cemented layers on steels were calculated for each temperature. The values of the $K_T$ constant are given in Table 3.

**Table 3.** Values of the kinetic growth rate constants $K_T$ of the cemented layer on the surface of the investigated steels at different temperatures of the carburization

| Temperature (°K) | | 1143 | 1173 | 1203 |
|---|---|---|---|---|
| $K_T x 10^7$ ($cm^2/s$) | 18CN4 | 1.80 | 0.97 | 0.62 |
| | 22MC4 | 2.84 | 2.01 | 1.25 |

The growth rate constant is related to the inverse of the treatment temperature according to the Arrhenius equation for thermally activated processes [13-15].

$$LnK = LnK_0 - Q/_{RT} \tag{2}$$

where $K_0$ is a pre-exponential constant, Q is the activation energy (J/mol.), T is the absolute temperature (in °K) and R is the gas constant (J/mol.K).

Figure 4 shows the variation of the growth rate constant $K_T$ of carburizing layers of steels as a function of the inverse temperature $T^{-1}$. Activation energies are calculated from the slopes of the plots and their values are 155.11 and 199.54 kJ/mol. for 22MC4 and 18CN4 steels respectively.

One should notice that the 22MC4 steel has an ability to be carburized in a shorter time than that for the other steel. This effect is plausibly due to an increased content of manganese in the 22MC4 steel. Addition of this element promotes in general the hardness of the steel, however it reduces toughness of the steel [16].

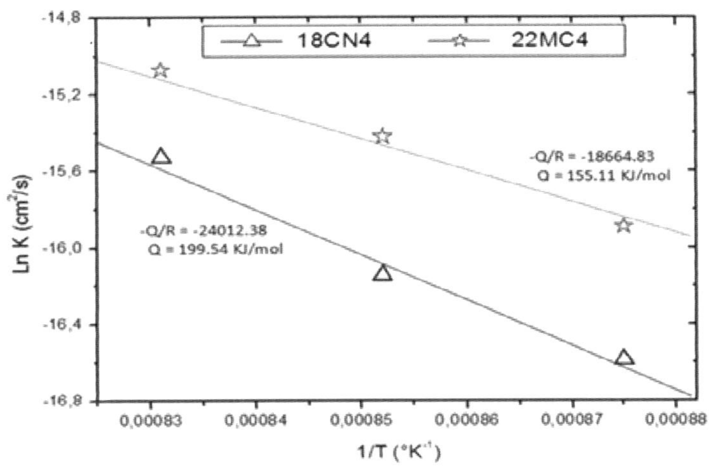


**Fig. 4** Growth rate constant $K_T$ of the thickness of cemented layer versus inverse temperature of carburizing steels 18CN4 and 22MC4

*3.4. Hardness*

Figure 5 shows the hardness profiles obtained for 18CN4 and 22 MC4 steels carburized for the time of 6 hours at temperature of 1203 °K. It is shown that the hardness profile curves have a similar shape with three distinct regions. The first region, corresponding, to the outer layer, is characteristic of the highest hardness. This feature can be explained by an important diffusion rate of carbon atoms to the austenite during soaking period. As a result, the martensite originated from the saturated austenite phase has the highest hardness. The second region which is called the diffusion zone, shows the descending hardness values which can be explained by a decreasing diffusion flow of carbon atoms with an increasing depth from the steel surface. As a result the phase composition in this region is changing from martensitic to semi-martensitic and next to ferrito-perlitic one. For the third region, the hardness has almost stationary value which corresponds to the intact core of the steel. The hardness values for the 22MC4 steel are slightly higher than for the 18CN4 one due to greater carbon and manganese content in the steel.

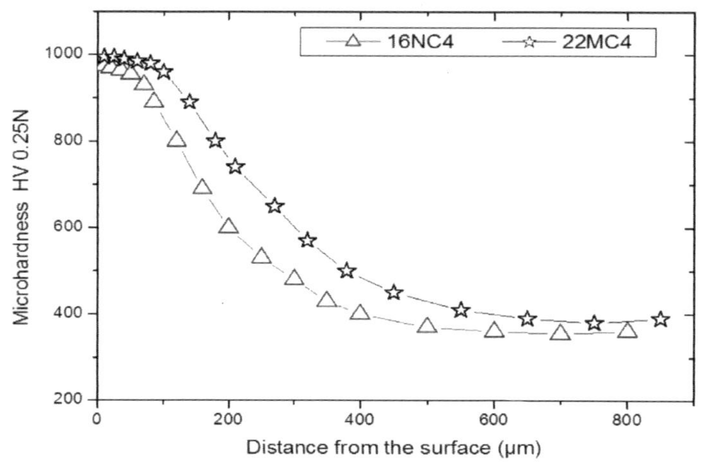


**Fig. 5** Hardness profiles for the both steels carburized for the length of time 4 h at 1203 °K

## 4. Conclusions

Powder pack carburizing treatment performed for two low alloy steel grades confirm possibility to optimize useful properties of the steels with use two experimental parameters as carburization temperature and the length of time of the soaking process.

## Acknowledgements

The present work was partly financially supported by the Thematic Agency Research of Science and Technology (ATRST) of Algeria under research contract of thematic project n° 100/2015.

## References

[1] Terheci M and Li S 1997 Stress-overstress-surface fatigue: new approach in the understanding of the Tribology of plasma coated sintered steels *Wear* **210** pp 17-26

[2] Li C, He Q, Tang W and Lu F 2004 Carburising of steel AISI 1010 by using a cathode arc plasma process *Surface & Coatings Technology* **187** pp 1-5

[3] Frandsen R B, Christiansen T and Somer M A J 2006 Simultaneous surface Engineering and bulk hardening of precipitation hardening stainless steel *Surface & Coatings Technology* **200** pp 5160-5169

[4] Yan M F and Liu Z R 2001 Study on microstructure and microhardness in surface layer of 20CrMnTi steel carburised at 880°C with and without RE *Materials Chemistry and Physics* **72** pp 97-100

[5] Sun Y and Haruman E 2008 Influence of processing conditions on structural characteristics of hybrid plasma surface alloyed austenitic stainless steel *Surface & Coatings Technology* **202** pp 4069-4075

[6] Schneider A 2002 Iron layer formation during cementite decomposition in carburising atmospheres *Corrosion Science* **44** pp 2353-2365

[7] Sun Y 2005 Hybrid plasma surface alloying of austenitic stainless steels with nitrogen and carbon *Materials Science and Engineering A* **404** pp 124-129

[8] Nart E and Genel K 2010 Investigation of carburising treatment of injector flange in diesel engine *Materials and Design* **31** pp 2110-2117

[9] Ohmori Y and Sugisawa S 1971 The precipitation of carbides during tempering of high Carbon martensite *Trans. Japan Inst. of Metals* **12** pp 170-172

[10] Scheuer C J, Cardoso R P, Mafra M and Brunatto S F 2013 AISI 420 martensitic stainless steel low temperature plasma assisted carburizing kinetics *Surf. Coat. Technol.* **214** pp 30-37

[11] Benarioua Y and Maza S 2015 Towards carburizing treatment applied on low alloy steel 16CN6 *Proceedings of Engineering and Technology - PET* Copyright - IPCO **9** Special issue 2nd International conference on Renewable Energy (CIER-2014)

[12] Gressmann T, Nikolussi M, Leineweber A and Mittemeijer E J 2006 Formation of massive cementite layers on iron by ferritic carburising in the additional presence of ammonia *Scripta Materialia* **55** pp 723-726

[13] Bell T and Sun Y 2002 Low-temperature plasma nitriding and carburising of austenitic stainless steels *Heat Treatment of Metals* **29** pp 57-64

[14] Sun Y 2005 Kinetics of low temperature plasma carburizing of austenitic stainless steels *Journal of Materials Processing Technology* **168** pp 189-194

[15] Chen F S and Wang K L 1999 The kinetics and mechanism of multi component diffusion on AISI 1045 steel *Surface & Coating Technology* **115** pp 239-248

[16] Suutala N 1982 Effect of manganese and nitrogen on the solidification mode in austenitic stainless steel welds *Metallurgical Transactions A*, **13** pp 2121-2132

# High content palladium nanocomposite carbon-palladium films

**Joanna Rymarczyk[1], Elżbieta Czerwosz[1], Piotr Dłużewski[2], Ryszard Diduszko[1], Mirosław Kozłowski[1]**

[1] Tele and Radio Research Institute, Ratuszowa 11, 03-450 Warsaw, Poland
[2] Institute of Physics PAS, Al. Lotników 32/46, 02-668 Warsaw, Poland

e-mail joanna.rymarczyk@itr.org.pl

**Abstract**. The results of studies on innovative high content palladium nanocomposite carbonaceous-palladium films (C-Pd film) obtained by Physical Vapor Deposition (PVD) method are presented. The influence of hydrogenation on these films' structure, morphology and electric properties were investigated. The studies of structure and morphology of such films by SEM, EDS, TEM, XRD, and FTIR methods are presented. Decrease of resistance of these films due to hydrogenation was found.

## 1. Introduction

Palladium acetate is one of the most popular precursors for preparing active palladium catalysts [1], because of its versatility in catalyzing C–H bond activation [2] and cross-coupling reactions [3] of many different types of organic molecules to make pharmaceuticals [4], various chemical products used in agriculture [5], and electronic materials [6]. Additionally palladium catalysts are used in many processes of the petrochemical and petroleum refining industries, including the hydrogenation of unsaturated compounds, dehydrogenation, isomerization and hydrogenolysis [7÷9].

A conventional way of preparing palladium catalysts consists of loading a support with the initial palladium compound followed by the reduction of this compound [10]. However the reduction mechanisms of the catalysts precursors on support surfaces are still not fully understood.

One of the widespread reductant in the synthesis of supported metal catalysts is hydrogen [11]. Palladium acetate reduction by hydrogen takes place according to the following overall chemical equation:

$$Pd(OAc)_2 + H_2 \rightarrow Pd + 2HOAc.$$

Palladium is also used in the preparation of metal complex forming coordination polymers. A polymer–metal complex is composed of synthetic polymer and metal ions bound to the polymer ligand by a coordinate bond [12].

Metal colloids have unique physical properties and high catalytic activity. A common method for obtaining palladium colloids is the reduction of palladium salts in the presence of a stabilizing agent, such as a polymer, which prevents the formation of catalytically inactive "palladium black" due to the existence of a sterile barrier between the nanoparticles. The palladium on the polymer is applied. Their high surface-to-volume ratios make metal colloids very good candidates for active catalysts, but their thermodynamic instability complicates practical applications [13].

Interest in palladium is also related to its hydrogen sorption properties. Palladium is one of the best adsorbents of hydrogen [14]. Molecular hydrogen is dissociated at the palladium surface into two H atoms before absorption. Hydrogen can move in a subsurface layer and diffuse into the metal host lattice as a solid solution (α-phase). As the hydrogen pressure together with its concentration in the metal increases, interactions between hydrogen become locally important, and start a nucleation and growth of the hydride, β-phase of metallic palladium hydride is formed [15÷20]. Palladium is an element which is capable of reversibly absorbing large amounts of hydrogen. Two H atoms can recombine to $H_2$ in the desorption process.

In the presence of $H_2$ the resistivity of Pd will change due to the formation of a solid solution of Pd/H [21÷23]. Furthermore, Pd is highly selective to $H_2$ [24].

In this paper we present new nanocomposite carbonaceous-palladium material (C-Pd film) with high content of palladium (more than 50 wt %) obtained by deposition of palladium acetate and $C_{60}$ from two separated sources in PVD process. We present results of studies on the structure changes of the C-Pd film due to the hydrogenation of film and on the other hand due to the reduction of the palladium acetate during technological process.

## 2. Experimental

### *2.1. Preparation of nanocomposite C-Pd films*
Nanocomposite C-Pd films were prepared by Physical Vapor Deposition (PVD) method. These processes were performed from two separate sources, one containing $C_{60}$ fullerenes ($C_{60}$, 99.9% Sigma-Aldrich) and other containing palladium acetate (Palladium (II) acetate, 99.98%, Sigma-Aldrich), both in the form of powder. The material is led to the evaporation by resistive heating. PVD process was conducted in a dynamic vacuum $10^{-3}$ Pa [25÷27]. The duration of the process was 5 min. Films were prepared on quartz glass polished substrates. The thickness of the films was 200÷300 nm.

### *2.2. Methods for characterization of nanocomposite C-Pd films*
The topography and morphology of the prepared C-Pd films were studied by Scanning Electron Microscopy (SEM) method and Energy Dispersive X-ray Spectroscopy (EDS) technique using JEOL JSM-7600F scanning electron microscope with INCA X-Max spectrometer.

Studies of crystalline structure of C-Pd films were per-formed using Transmission Electron Microscopy (TEM), electron diffraction of selected area (SAED). The Titan Cubed transmission electron microscope operating at 300kV accelerating voltage was used for high-resolution transmission electron microscopy (HRTEM).

X-ray diffraction (XRD) studies were performed with Rigaku SmartLab 3kW diffractometer at room temperature in the θ/2θ scanning mode with CuKα radiation and Si:Li semiconductor detector. Diffraction measurements were performed in grazing incidence primary beam geometry (GIXD).

The molecular structure of C-Pd films was investigated by infrared spectroscopy with Fourier transformation (FTIR). FTIR spectra were obtained with ThermoScientific Nicolet iS10 FTIR spectrometer, using ATR (Attenuated Total Reflection) technique, in the spectral range 500 ÷ 4000 $cm^{-1}$, with a resolution of 4 $cm^{-1}$.

The hydrogen sensing characteristics of C-Pd films in hydrogen gas presence were measured in a special set-up, designed and built in the Tele and Radio Research Institute [28]. All measurements were performed for the gas mixture of $H_2$ in nitrogen, with a rate of 50 ml/min. Electrical measurements were carried out in cycles of 10 min. In the first cycle, the gas with hydrogen was introduced to the experimental chamber. After 10 min, the synthetic air was introduced to remove $H_2$ gas. In the next cycle hydrogen gas was used again. Several hydrogen absorption-desorption cycles were carried out. Measurements were performed at room temperature and at atmospheric pressure.

The scheme of experimental set-up is shown in Fig. 1.

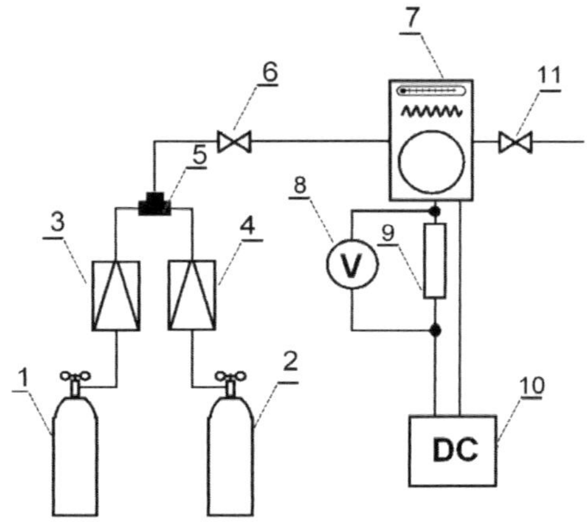


**Figure 1.** Experimental set-up used for hydrogen sensing measurements (1, 2 - gas bottles, 3, 4 - mass flow controllers, 5 – gas mixer, 6, 11 - valves, 7 - measurement chamber, 8 – voltmeter, 9 - reference resistor, 10 - voltage source)

The response of C-Pd films to hydrogen is defined as the percentage change in film resistance and was calculated using the following equation:

$$\Delta R/R_0 = ((R - R_0)/R_0) \cdot 100\%,$$

where R - film resistance after exposure to $H_2$, $R_0$ - film initial resistance in $H_2$ absence.

The electrodes for electrical measurements were prepared of silver paste. Two electrodes were placed on the film surface at a distance of 1 cm from each other. Estimated contact resistance electrode-probe was ~ 0.1 Ω while films resistance was higher than 100 Ω.

## 3. Results and discussion

### *3.1. Characterization of nanocomposite C-Pd films*

Nanocomposite C-Pd films are composed of palladium nanograins embedded in various carbonaceous matrixes and are characterized by high content of palladium, above even 50 wt %. EDS spectrum of the C-Pd film is show in Fig. 2. Observation of topography by SEM method indicated that nanocomposite C-Pd films have very smooth surface. Exemplary SEM image of C-Pd film surface topography is presented in Fig.3.

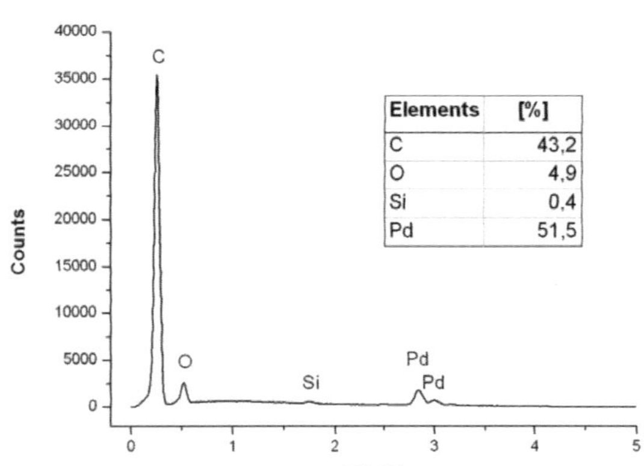


**Figure 2.** EDS analysis of the C-Pd film

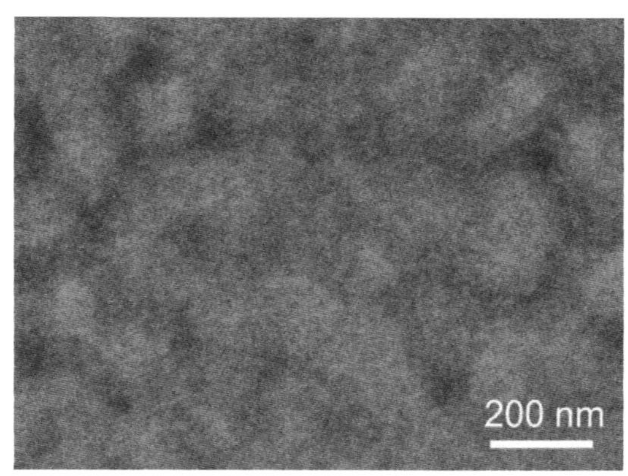


**Figure 3.** SEM image of the topography of nanocomposite C-Pd film

TEM study revealed that nanocomposite C-Pd films are composed of palladium nanocrystals and carbonaceous matrix. TEM image of the C-Pd film is shown in Fig.4. Pd nanograins, seen in TEM image as black objects, are very small with diameter of few nm. The Pd nanograins are randomly distributed over the entire carbonaceous matrix. The histogram of C-Pd film is shown in Fig. 4b. The average grain size (rs) of Pd nanograins determined from the histogram in this film is 2.5 nm and all observed objects have diameters smaller than 4 nm.

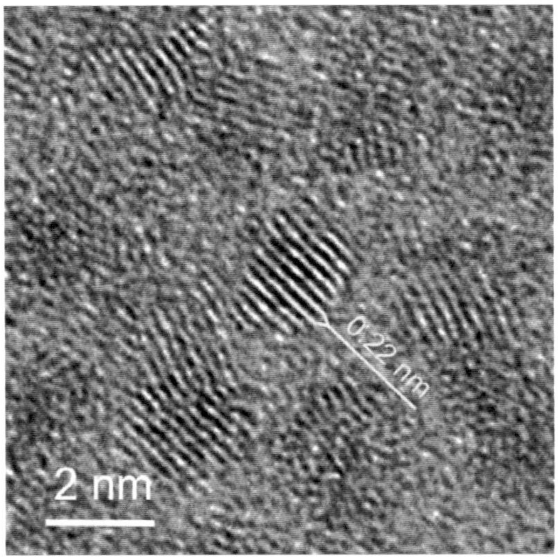


**Figure 4a.** TEM image of the nanocomposite C-Pd film; b) histogram of Pd nanograins size of C-Pd film

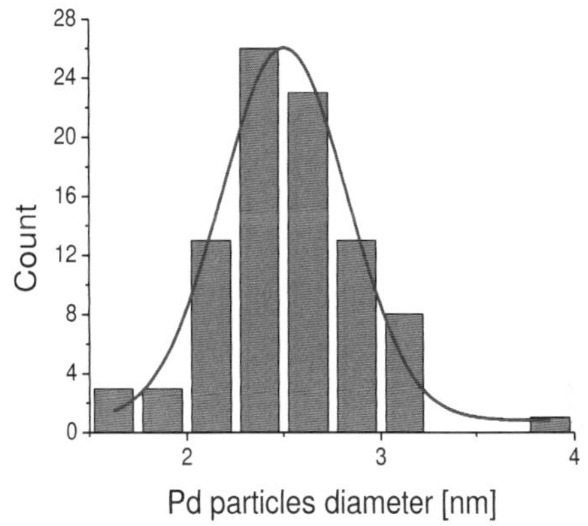


**Figure 4b.** Histogram of Pd nanograins size of C-Pd film

In Fig. 5 fast Fourier transformation (FFT) diffraction patterns from presented fragment in Fig. 3a of C-Pd film is shown. Visible in Fig. 5 diffused ring could be attributed to (*hkl*) planes of Pd grains with the structure of *fcc* type (marked with yellow circle in Fig. 5). The diffused rings prove the presence of numerous small grains in the C-Pd film.

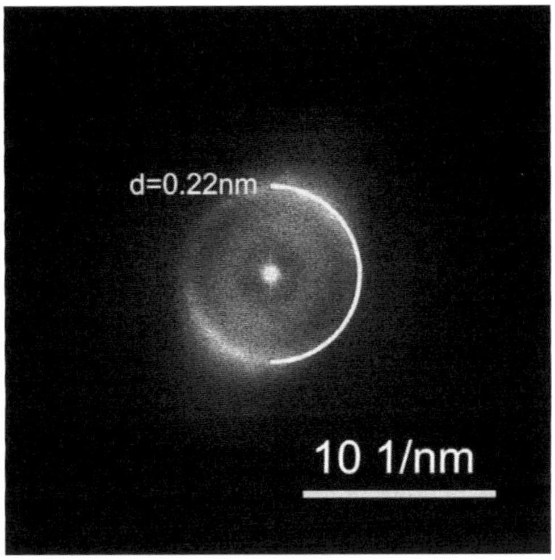


**Figure 5.** FFT diffraction patterns from the nanocomposite C-Pd film

### *3.2. The hydrogen sensing characteristics of C-Pd films*

The hydrogen response real time characteristic for nanocomposite C-Pd film is presented in Fig. 6. In the first contact with hydrogen, C-Pd films' resistance decreases (blue line in Fig. 6). As a result of such activation, the resistance decreased from 31 MΩ to 1.7 kΩ. Then, the experiment of hydrogenation in several cycles was performed. The big decrease of resistance after first hydrogenation can be connected to reduction of palladium acetate to crystalline form of metallic palladium grains by hydrogen. The introduction of synthetic air in every cycle did not cause appreciable return of resistance to the initial value. In the next measurement cycle (red line in Fig. 6) the resistance permanently increases and it stabilizes at a certain level after many cycles realization. It is connected with solution of hydrogen in palladium nanograins and the formation of solid solution of $PdH_x$ (resistance increase) and decomposed to Pd nanograin in air (resistance decrease).

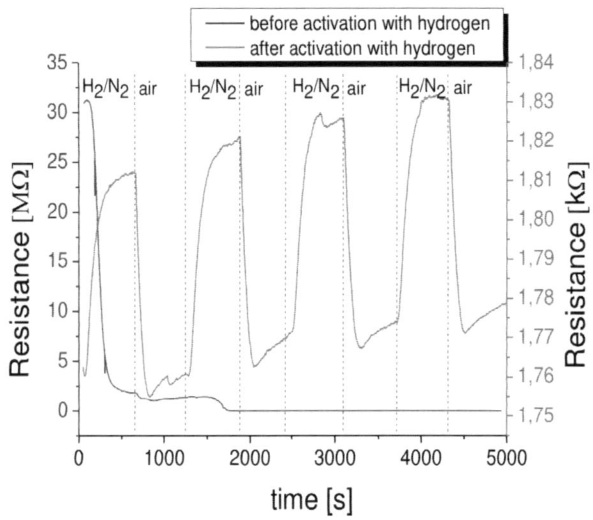


**Figure 6.** Hydrogen response characteristics for the nanocomposite C-Pd film

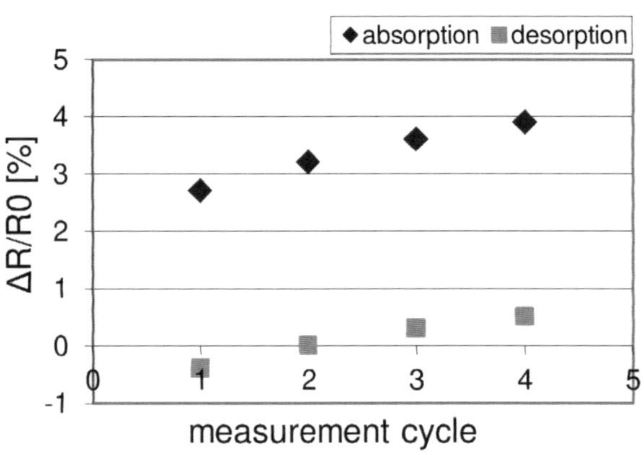


**Figure 7.** Hydrogen response of C-Pd film as percentage change in film resistance in subsequent measurement cycles

The response of C-Pd film to hydrogen as the percentage change in film resistance is shown in Fig. 7. In this graph we can see that the initial resistance to the reaction with hydrogen increases by 2.7 % in first cycle but in the last by 3.9 %. The $\Delta R/R_0$ increase in the subsequent measurement cycles is noticeable. The cause is an incomplete decrease in resistance upon introducing nitrogen into a measurement chamber. After the first cycle, the resistance decreased below the initial value. The reintroduction of cyclic mixture of hydrogen and synthetic air caused an increase of the resistance when hydrogen gas was introduced and a decrease when synthetic air was introduced. However, resistance did not return to the initial value in the synthetic air within 10 minutes what results in an increased sensitivity in the following cycles of measurement.

### *3.3. Characterization of nanocomposite C-Pd films after activation*

SEM image of the topography of the surface of nanocomposite C-Pd film after hydrogen measurement is presented in Fig. 8. Comparing these results with the SEM image before the hydrogen sensing measurement (Fig. 3), no changes were observed.

**Figure 8.** SEM image of the nanocomposite C-Pd film after hydrogen measurement

However, changes in the crystalline structure of nanocomposite C-Pd films were noticed. In TEM images of the C-Pd film after hydrogen measurement (Fig. 9), we observed increase in grain's diameter while comparing with Pd grains observed in Fig. 4. Additionally a large dispersion of the distribution of the size of nanograins of C-Pd films is visible. In Fig. 9a Pd grains with different diameters are found. Many small nanograins with diameters to 10 nm are visible, but the hydrogen measurement caused also migration and agglomeration of palladium to big palladium particles. In the volume of the C-Pd film, the joined chain forming particles are observed. These chains can create current conduction paths. That could explain the reason of the C-Pd film's resistance decrease after the hydrogen measurement.

The histogram of nanograins diameter distribution for sample after hydrogen measurement (Fig. 9b) included only smaller grains because it is difficult to distinguish individual grains in the agglomerate. The histogram shows that the average grain size (rs) of Pd nanograins is 6 nm. This means that due to the hydrogen measurement, the grain's diameter increased twice. There were also differences in Fast Fourier Transformation (FFT) diffraction patterns from C-Pd before and after hydrogen measurement (Fig. 10).

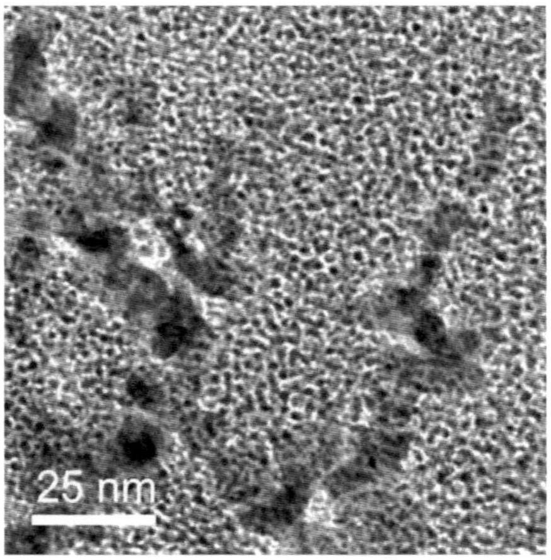


**Figure 9a.** TEM image of the nanocomposite C-Pd film after hydrogen measurement

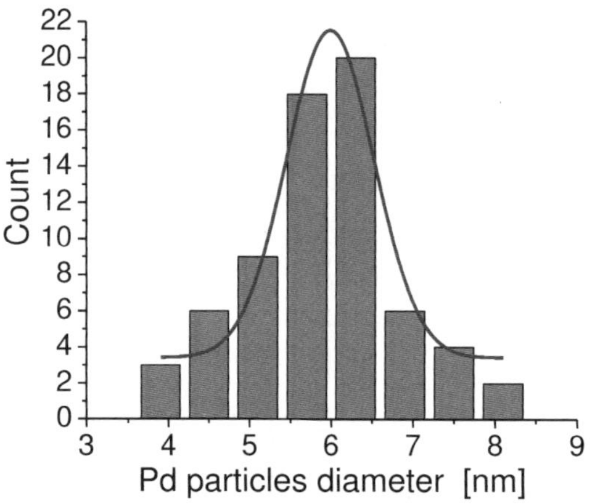


**Figure 9b.** Histogram of Pd nanograins size of C-Pd film film after hydrogen measurement

Large dispersion of the size distribution of Pd nanograins reflects in the electron diffraction patterns (Fig. 10) where both diffuse diffraction rings (for small crystallites) and bright spots (for large crystallites) are visible. The first two inner rings, corresponding to distances between {111} and {002} palladium crystal are found.

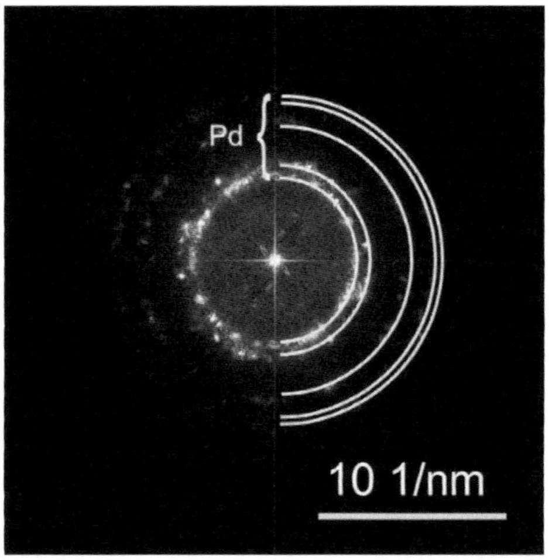


**Figure 10.** Fast Fourier Transformation (FFT) diffraction pattern from the nanocomposite C-Pd film

X-ray diffraction studies have shown a change in the structure of the C-Pd film due to hydrogenation. In XRD pattern for initial C-Pd film (obtained by PVD process) we observe palladium acetate (Fig. 11 black line) while crystalline form of metallic palladium grains is no visible. XRD spectrum of C-Pd film (red line) after hydrogenation demonstrates that peaks of metallic palladium are observed. The Pd peak's intensity indicates a small size of these grains. XRD results suggest a formation of Pd nanograins after hydrogenation of films and disappearing of palladium acetate crystalline form.

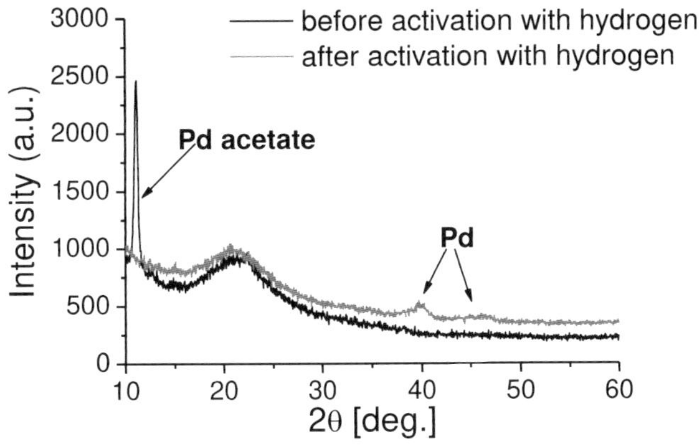


**Figure 11.** XRD spectra of the C-Pd film after PVD process – black line and after hydrogen measurement – red line

The absorption spectra for infrared region for nanocomposite C-Pd film are presented in Fig. 12 (black line). In the region 400 ÷ 2000 $cm^{-1}$ the characteristic band for $C_{60}$ molecule vibration (1182 $cm^{-1}$) and characteristic bands of some characteristic groups for palladium acetate $Pd(CH_3COO)_2$ are found. The bands attributed to vibrations of palladium acetate are as follows: asymmetric stretching vibrations of COO at 1605 $cm^{-1}$; band attributed to a symmetric stretching vibrations COO at 1432 $cm^{-1}$ and band characteristic for $CH_3$ bending vibrations - 1335 and 695 $cm^{-1}$ [29÷32]. FTIR spectrum for hydrogenated film does not exhibit characteristic bands of palladium acetate (red line on Fig. 12). This result confirm conclusion from XRD studies about decomposition of palladium acetate and modification of Pd crystalline structure to the metallic.

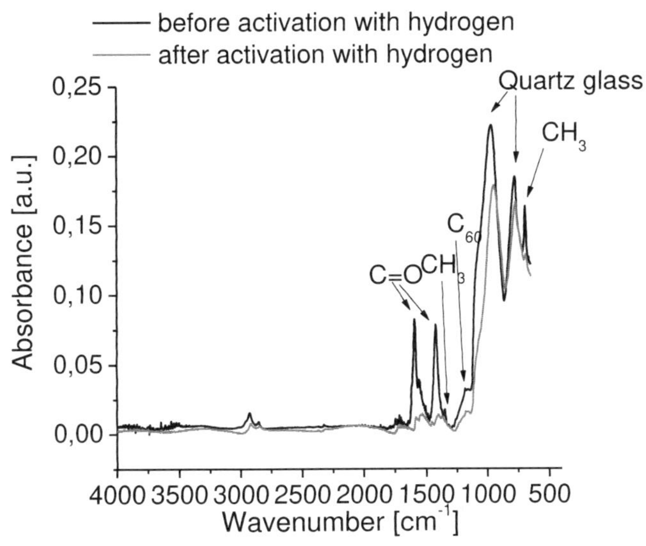


**Figure 12.** FTIR spectra of the C-Pd film after PVD process – black line and after hydrogen measurement – red line

## 4. Conclusions

Concluding, the results of XRD, FTIR and hydrogenation experiments show that C-Pd films with high content of palladium after first hydrogenation process change their form (palladium acetate disappears while Pd nanograins are formed). Due to this reorganisation of the film structure, the stable film containing palladium nanograins with average diameter of 6 nm is formed.

During many cycles of the gassing/degassing process the resistance of C-Pd films permanently changes. The effect can be connected to the hydrogen adsorption/desorption and the formation of $PdH_x$ nanograins during the adsorption process. During the desorption process $PdH_x$ grains turn into Pd grains, but not completely. Due to the big size (6nm in diameter) of Pd nanograins, hydrogen desorption is not total and the resistance of film increases. On the other hand the formation of conductivity paths consisted of Pd nanograins is observed.

These structural changes in the C-Pd films are stable.

**Acknowledgements**
This work was co-financed by the European Regional Development Fund within the Innovative Economy Operational Programme 2007-2013 No UDA-POIG.01.03.01-14-071/08-00 and Ministry of Science and Higher Education under Research Project “Nanocomposite C-Ni-Pd films for opto- and bio- electronics applications”.

**References**

[1] Carole WA, Colacot TJ 2016 *Chem. Eur.J.* 2016, **22**, 7686
[2] Zaitsev VG, Shabashov D, Daugulis O 2005 *J. Am. Chem. Soc.* **127** 13154
[3] Miyaura N, Suzuki A. 1995 *Chem. Rev.* **95** 2457
[4] Doucet H, Hierso JC 2007 *Curr. Opin. Drug Discov. Devel.* **10** 672
[5] Torborg C and Beller M 2009 *Adv. Synth. Catal.* **351** 3027
[6] Bracher C, Yi H, Scarratt N, Masters R, Pearson A, Rodenburg C, Iraqi A., Lidzey D 2015 *Org. Electron.* **27** 266
[7] Ravanchi MT, Kaghazchi T, Kargari A 2009 *Desalination* **235** 199
[8] Stern EW, Maples PK 1972 J. Catal. **27** 120
[9] Brandt B, Fischer J-H, Ludwig W, Libuda J, Zaera F, Schauermann S and Freund H-J 2008 *J. Phys. Chem. C* **112** 11408
[10] Corral JA, López MI, Esquivel D, Mora M, Jiménez-Sanchidrián C and Romero-Salguero FJ 2013 *Materials* **6** 1554
[11] Berenblyum AS, Al-Wadhaf HA, Katsman EA and Flid VR 2011 *Kinet. Catal.* **52** 296
[12] Kaliyappan T, Kannan P 2000 *Prog. Polym. Sci.* **25** 343
[13] Gniewek A, Trzeciak AM, Ziółkowski JJ, Kępiński L, Wrzyszcz J, Tylus W 2005 *J. Catal.* **229** 332
[14] Lewis FA 1982 *Platinum Metals Rev.* **26** 20
[15] Johansson M, Skúlason E, Nielsen G, Murphy S, Nielsen RM, Chorkendorff I 2010 *Surf. Sci.* **604** 718
[16] Ramanathan M, Skudlarek G, Wang HH, Darling SB 2010 *Nanotechnology* **21** 125501
[17] Matsumura D, Okajima Y, Nishihata Y, Mizuki J 2011 *J. Alloy. Compd.* **509** S849
[18] Borgese L, Zanola P, Bontempi E,Rossi D, Depero LE 2010 *J. Coat. Technol. Res.* **7** 691
[19] Maeland A, Flanagan TB 1966 *Platinum Metals Rev.* **10** 20
[20] Züttel A 2007 *Mitig. Adapt. Strat. Glob. Change.* **12** 323
[21] Kamińska A, Diduszko R, Czerwosz E, Kowalska E, Kozłowski M 2014 *Int J Hydrogen Energ* **39** 9854
[22] Kamińska A, Krawczyk S, Kozłowski M, Czerwosz E, Sobczak K. 2013 *Sens. Lett.* **11** 500
[23] Kowalska E, Czerwosz E, Diduszko R, Kaminska A, Danila M 2013 *Sensor. Actuator. A* **203** 434
[24] Kamińska A, Krawczyk S, Czerwosz E, Sobczak K, Kozłowski M 2013 *Sensor. Actuator. A* **196** 86
[25] Rymarczyk J, Czerwosz E, Kozłowski M, Dłużewski P, Kowalski W 2014 *Pol. J. Chem. Technol.* **16** 18
[26] Sobczak K, Dłużewski P, Klepka MT, Kurowska B, Czerwosz E 2012 *Int. J. Hydrogen Energy* **37** 18556

[27] Czerwosz E, Dłużewski P, Kowalska E, Kozłowski M, Rymarczyk J 2011 *Phys. Status Solidi C* **7-8** 2527
[28] Czerwosz E, Dłużewski P, Kozłowski M, Krawczyk S, Rymarczyk J 2016 *Vacuum* **128** 265
[29] Kragten DD, van Santen RA 1999 *Inorg. Chem.* **38** 331
[30] Kuzmany H, Winkler R, Pichler T 1995 *Phys. Condens. Master* **7** 6601
[31] Rymarczyk J, Kamińska A, Kęczkowska J, Kozłowski M, Czerwosz E 2013 *Opt. Appl.* **43** 123
[32] Yi JY, Bernholc J 2005 *Chem. Phys. Lett.* **403** 359

IOP Conf. Series: Journal of Physics: Conf. Series **1033** (2018) 012010 doi:10.1088/1742-6596/1033/1/012010

# Conversion treatment of thin titanium layer deposited on carbon steel

**Younes Benarioua[1,4,*], Bogdan Wendler[2], Didier Chicot[3]**

[1] Département de Génie Mécanique, Faculté de Technologies, Université de M'sila, Route de Bordj Bou Arréridj - M'sila 28000, Algérie.
[2] Institute of Materials Science and Engineering; Lodz University of Technology; 60- 924 Lodz, Poland.
[3] Université Lille Nord de France, Lille1, LML, UMR 8107, F-59650 Villeneuve d'Ascq, France.
[4] Unité de Recherche Matériaux Emergents, Université de Sétif 1- 19000, Algérie

* Corresponding author e-mail: benariouayounes@yahoo.fr

**Abstract.** The present study has been conducted in order to obtain titanium carbide layer using a conversion treatment consisting of two main steps. In the first step a thin pure titanium layer was deposited on 120C4 carbon steel by PVD. In the second step, the carbon atoms from the substrate diffuse to the titanium coating due to a vacuum annealing treatment and the Ti coating transforms into titanium carbide. Depending on the annealing temperature a partial or complete conversion into TiC is obtained. The hardness of the layer can be expected to differ depending on the processing temperatures. By a systematic study of the hardness as a function of the applied load, we confirm the process of growth of the layer.

## 1. Introduction

The use of transition metal carbides and nitrides as coating materials to improve the workpieces surfaces has shown an extensive development in the last years. Among these metallic compounds, titanium carbide presents excellent mechanical and chemical properties as like a high melting point, strong hardness and great chemical and thermal stability under severe conditions [1-7]. Titanium carbide coatings produced by PVD methods have a high hardness and resistance to wear and corrosion but have a low adherence to their substrates [8]. This last behaviour was improved by conversion treatment of titanium and chromium films deposited respectively on carbon steel and cemented steel [9-11].

This work is to examine then how the precipitation of the carbide layer, as a function of the treatment temperature, influences the measurement of hardness under different loads. The deposited titanium film can be transformed into titanium carbide partially or wholly. Several characterization methods show that the growth of the titanium carbide layer takes place from the substrate-film interface towards the coating's surface.

## 2. Experimental details

The material of substrates used for this study is 120C4 (AFNOR) steel. The chemical composition in weight is 1.20 %C, 0.11 % Si, 0.21 % Mn, 1% Cr, 0.2 % Ni, 0.01 % Al and 0.09 Cu. The cylindrical samples (diameter 15 mm, height 15 mm) were specularly polished before ionic cleaning and then a pure of thin titanium film (4 μm) was deposited by magnetron sputtering under argon atmosphere and a pressure of 3.10$^{-3}$ mbar. A description of the deposition technique of titanium films is given in [11]. In order to study the conversion process of titanium film into a titanium carbide one, coated simples were annealed in a tubular furnace at a vacuum of 1.5 10$^{-5}$ mbar during 1 h at temperature ranging from 500 to 1100°C.

The structures of films were analyzed using a HZG-3 Bragg–Brentano diffractometer with a monochromatic Co (Kα) radiation. Microstructure of samples (layers/substrates) was performed using an OM of type Olympus B O71 and SEM of type JEOL JSM 35. Further observations were performed on slope cut specimen for which the inclination angle relative to their surface was 2.5°. The samples were polished and etched in a solution composed of 100 ml of water, 2 ml of HF at 40% and 5 ml of $H_2O_2$ at 30%. Chemical composition throughout the films and substrates was evaluated using Energy-Dispersive Spectroscopy (EDS). Vickers indentations for loads ranging from 0.1 to 10 N have been applied using a Leco microhardness tester. At least three indentations were performed for each load in order to obtain a significant statistical value of hardness.

## 3. Results and discussion

### *3.1. Structure*

Fig. 1 shows the X-ray diffraction patterns before and after annealing of titanium deposited on 120C4. It should be noted that for the sample annealed at 500°C, the film presents only titanium in the form of Ti-α. The precipitation of titanium carbide starts at 700°C. The progression of the transformation process increases with the temperature from 700 to 1100°C. The transformation is completed for Ti deposited on the steel at 1100°C.

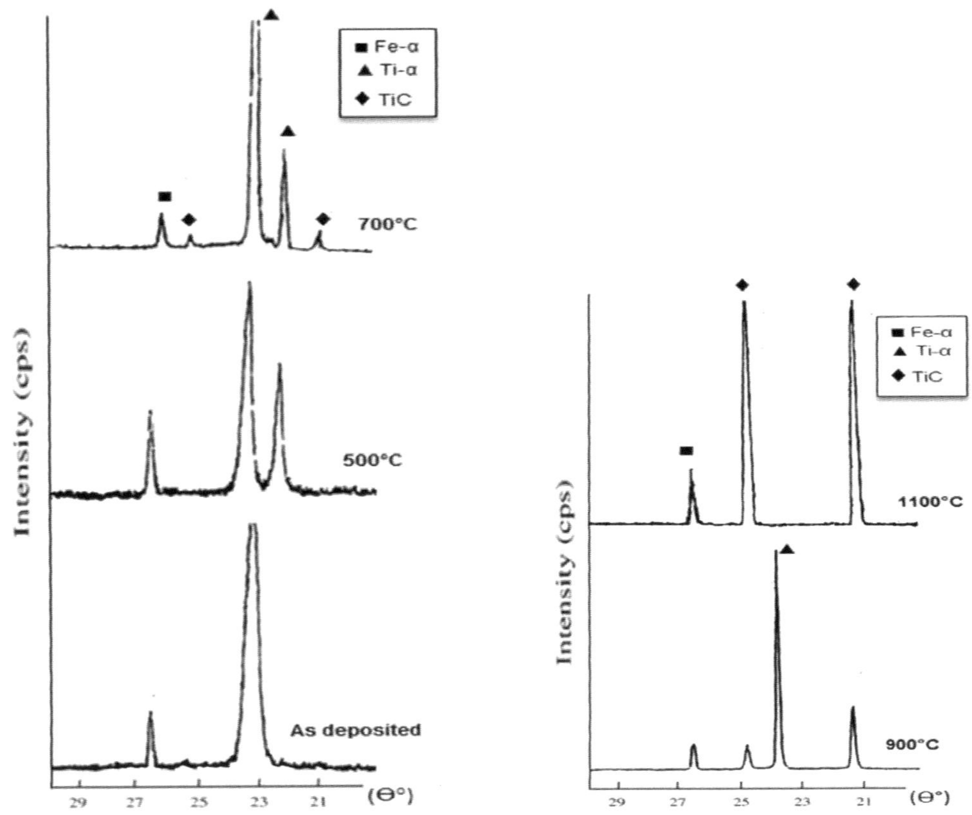


**Fig. 1**: X-ray diffraction spectra of titanium film deposited on 120C4 and annealed at various temperatures.

*3.2. Mirostructure*

The microstructure indicated in fig. 2A, B and C clearly shows the conversion of the film. At 700°C, the titanium carbide is visible as the edge at the film/substrate interface. The precipitation begins at the interface and propagates toward the surface. At 1100°C titanium film is totally transformed in titanium carbide. As indicated by scanning electron microscopy, the transformation of globular cementite into lamellar pearlite begins just then at the interface for low temperature and propagates towards the core of the substrate when the annealing temperature increases.

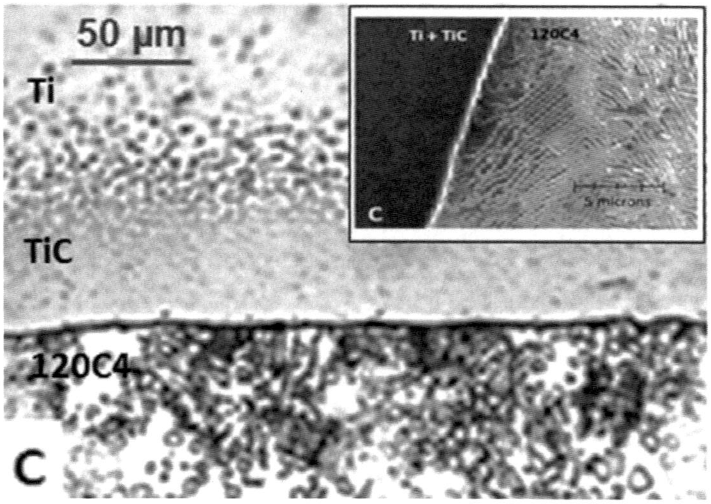


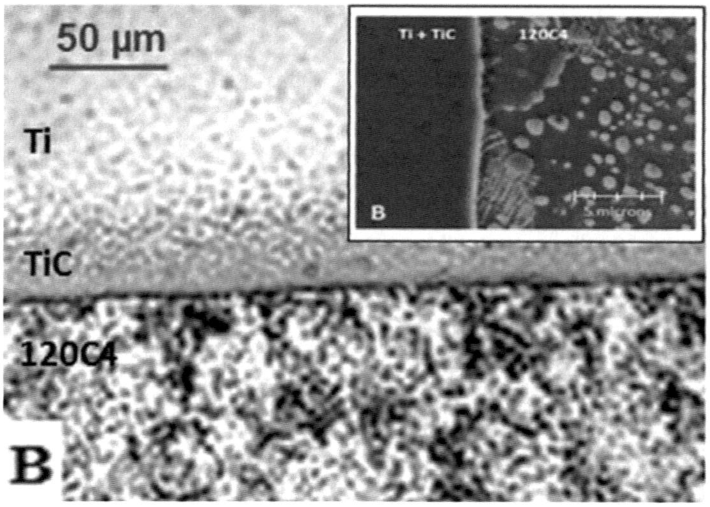

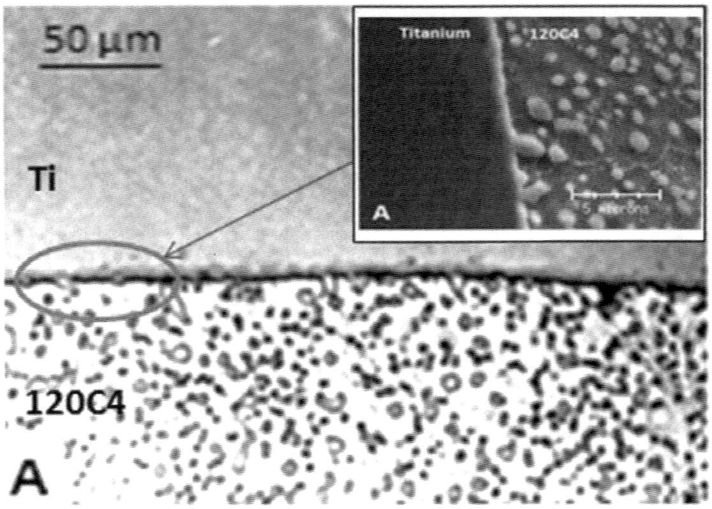


**Fig. 2:** OM and SEM images of Ti deposited on 120C4 steel after vacuum annealing at temperature 500°C (A), 700°C (B) and 900°C (C).

Fig. 3A and B present the concentration profiles of iron, chromium, titanium and carbon elements obtained by Energy Dispersive Spectrometry of samples annealed respectively at 700 and 1100°C. Carbon diffusion starts to be visible in the sample treated at 700 °C. The gradients Ti and Fe at the film–substrate interface indicated in fig.3A can be considered as artifacts due to lateral resolution intrinsic limitation of the EDS technique. At 1100°C, the analysis demonstrates that the transformation of Ti in TiC is homogeneous over the entire film. The precipitation of titanium carbide occurred can be explained by a large amount of carbon atoms diffused from substrate into titanium layer.

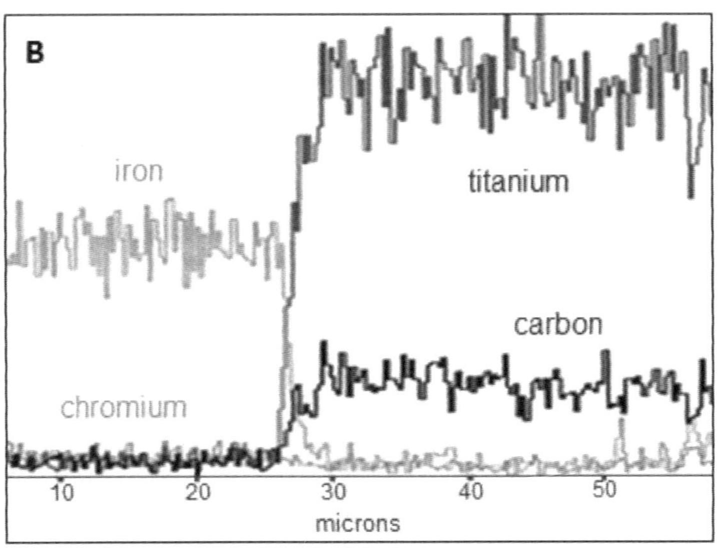

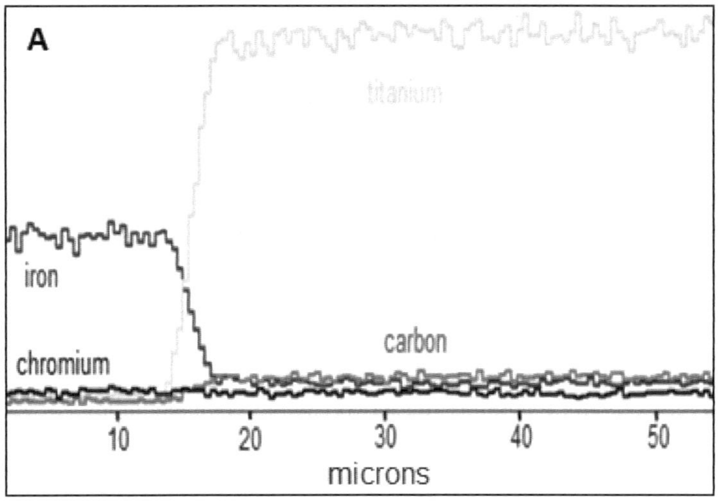


**Fig. 3:** Concentration profiles of the elements Fe, Ti, Cr and C obtained by EDS of samples annealed at 700 and 1100°C.

### *3.3. Hardness*

Table 1 collects the results obtained for indentations at the surface of the films and also the results of indentations performed on the substrate. We notice that there is a significant increase in hardness measured in film faces with increase of annealing temperature. For the substrates, the results are similar at 500, 700 and 900 °C whereas at 1100 °C the hardness was increased.

**Table 1:** Results of the Vickers hardness tests (mean of three indentations for each load)

| T(°C) | | P (Kgf) | 0.1 | 0.25 | 0.50 | 1 | 2 | 3 | 5 | 10 |
|---|---|---|---|---|---|---|---|---|---|---|
| As deposited | Film | d (μm) | 6.37 | 10.75 | 16.08 | 24.53 | 36.65 | 46.55 | 62.13 | 89.07 |
| | | H (GPa) | 4.54 | 4.01 | 3.58 | 3.08 | 2.76 | 2.57 | 2.40 | 2.37 |
| | Subs. | d (μm) | 9.45 | 15.15 | 21.30 | 30.03 | 42.61 | 52.60 | 68.06 | 96.61 |
| | | H (GPa) | 2.07 | 2.02 | 2.04 | 2.06 | 2.05 | 2.01 | 2.00 | 1.99 |
| 500 | Film | d (μm) | 6.43 | 11.41 | 17.46 | 26.73 | 39.45 | 49.52 | 65.07 | 93.61 |
| | | H (GPa) | 4.49 | 3.76 | 3.04 | 2.60 | 2.39 | 2.27 | 2.19 | 2.12 |
| | Subs. | d (μm) | 9.26 | 15.07 | 21.56 | 30.64 | 43.23 | 53.34 | 69.30 | 97.23 |
| | | H (GPa) | 2.16 | 2.04 | 1.99 | 1.98 | 1.98 | 1.96 | 1.93 | 1.95 |
| 700 | Film | d (μm) | 5.75 | 10.09 | 15.88 | 24.50 | 37.34 | 47.95 | 63.81 | 92.72 |
| | | H (GPa) | 5.62 | 4.75 | 3.67 | 3.09 | 2.64 | 2.43 | 2.28 | 2.16 |
| | Subs. | d (μm) | 9.46 | 14.90 | 21.14 | 29.79 | 42.28 | 51.72 | 67.69 | 95.56 |
| | | H (GPa) | 2.07 | 2.09 | 2.10 | 2.09 | 2.08 | 2.02 | 2.02 | 2.03 |
| 900 | Film | d (μm) | - | 7.38 | 11.84 | 18.38 | 28.44 | 37.75 | 53.69 | 78.45 |
| | | H (GPa) | - | 8.52 | 6.61 | 5.49 | 4.59 | 3.91 | 3.22 | 3.01 |
| | Subs. | d (μm) | 8.08 | 13.35 | 18.86 | 26.22 | 36.97 | 45.76 | 57.85 | 81.04 |
| | | H (GPa) | 2.84 | 2.60 | 2.61 | 2.70 | 2.72 | 2.66 | 2.77 | 2.82 |
| 1100 | Film | d (μm) | - | - | 7.35 | 11.61 | 17.92 | 22.67 | 30.68 | 44.96 |
| | | H (GPa) | - | - | 17.15 | 13.74 | 11.53 | 10.82 | 9.85 | 8.91 |
| | Subs. | d (μm) | 5.30 | 8.44 | 12.01 | 17.45 | 24.39 | 29.44 | 38.37 | 53.00 |
| | | H (GPa) | 6.62 | 6.51 | 6.43 | 6.10 | 6.24 | 6.42 | 6.30 | 6.60 |

In order to visualize the evolution of the hardness with the applied load, it is convenient to use the Meyer's relation which links the load P as a function of the impression diagonal d as [12].

$$P = a_m\, d^n$$

Where $a_m$ and n are linked by the dimensional equation and $n$ represents the hardness load dependence.

The use of this relation is not to obtain an intrinsic hardness but to study the annealing effect on the transformation of the film. The results of the tests samples indicated in table 1 can be presented using the relation of Mayer which is only valid in a load domain limited. For this, the resulting bilogarithmic diagrams take linear forms with $a_m$ and $n$ are coefficients. Generally the factor n can take three values. When $n = 2$, the hardness is independent of the load. If n takes values less than 2 ($n < 2$), the hardness increases with the load decreases. In the opposite case ($n > 2$), the composite hardness has to tend to substrate hardness when P tends to infinite values.

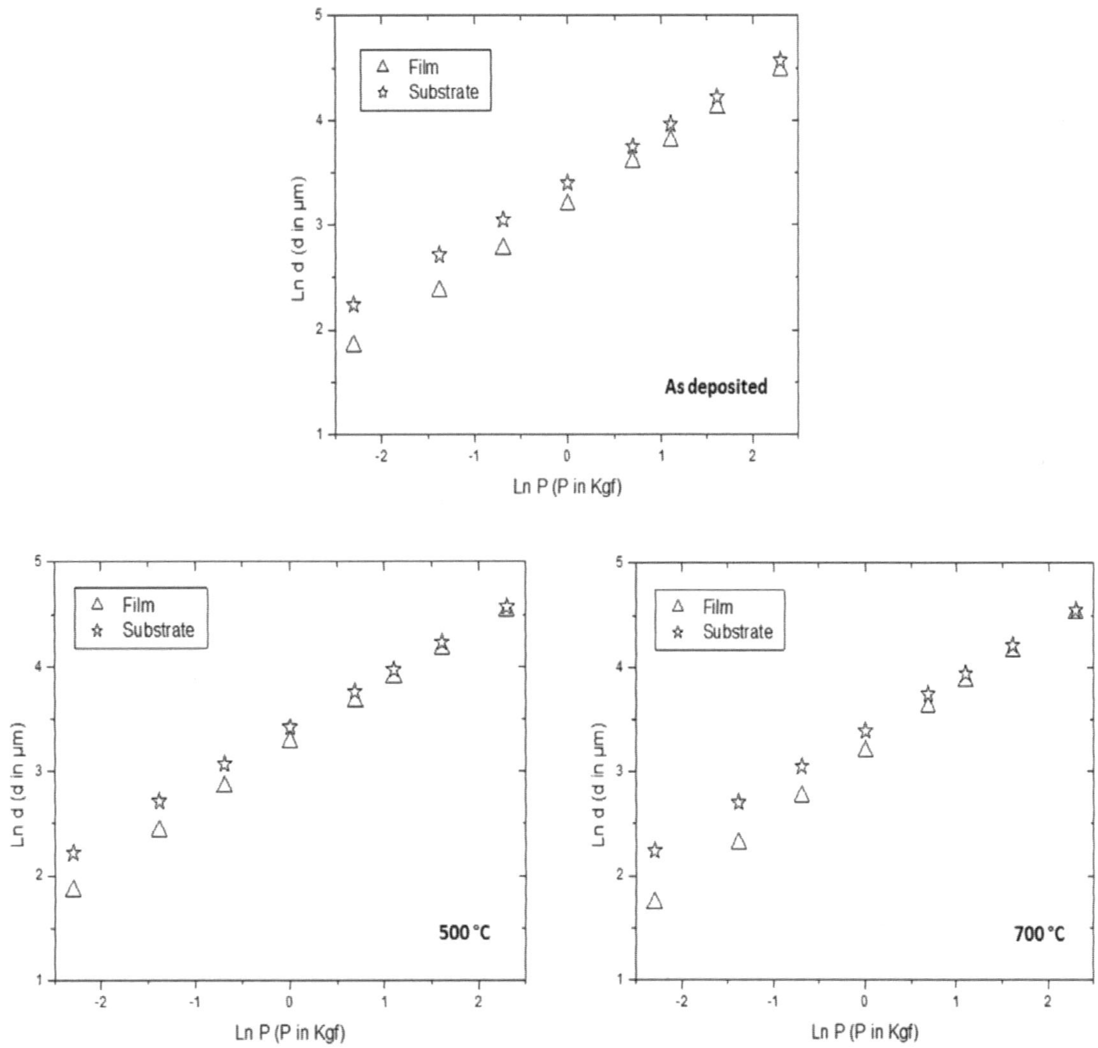

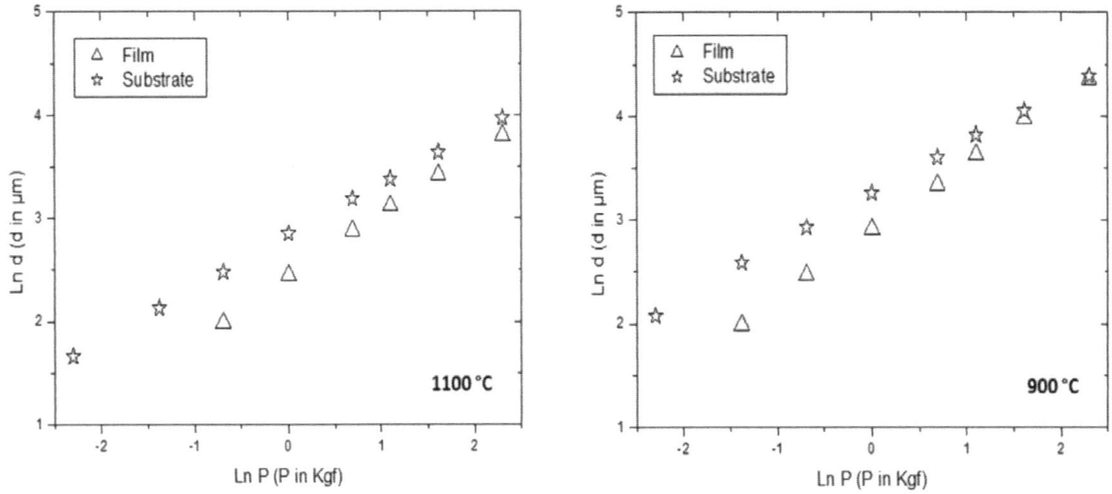


**Fig. 4:** Results of the indentation test of various samples according to the Meyer relation

We notice that as indicated in the fig. 4, the representation due to Meyer in our case doesn't show a slope break. A study of thick coatings (100 - 400 µm) of chromium carbide deposited on different substrates has shown that the representation of Mayer shows a break in slope [13].

**Table 2:** Values of Mayer's coefficient $a_m$ and $n$ of different samples

| Material type | Mayer Coeff. | As deposited | 500°C | 700°C | 900°C | 1100°C |
|---|---|---|---|---|---|---|
| Subst. | $a_m$ | 0.121 | 0.129 | 0.121 | 0.159 | 0.367 |
| | $n$ | 1.97 | 1.96 | 1.97 | 1.99 | 1.99 |
| Film | $a_m$ | 0.392 | 0.407 | 0.565 | 1.111 | 1.871 |
| | $n$ | 1.75 | 1.72 | 1.65 | 1.56 | 1.65 |

We will discuss the values of the coefficients $n$ and $a_m$ of the samples presented in tab. 2. For coefficient of films ($n \sim 1.70$) which is less than 2, the hardness measured on the surface of the films increases with the indentation load decreases. For the different charges applied to the substrates, the hardness obtained is constant. This is confirmed by $n$ corresponding to the substrates which equals 2. For the value of $a_m$, it remains constant when there is no transformation up to 700 ° C for the substrate and up to 500 ° C for the film. For the latter, progresses with the temperature in agreement the progression of the hardness and the metallographic observations on the one hand and X-ray diffraction on the other hand.

## 4. Conclusion

For titanium films deposited by magnetron sputtering on high carbon steel and for different annealing treatments, we have shown that the titanium is transformed into titanium carbide. The growth kinetics of titanium carbide from the substrate / film interface is strongly dependent on temperature. For a treatment time of one hour, we show that the titanium is completely transformed at 1100°C while at 500°C the transformation doesn't take place. The representation of Meyer, which joins the load applied to the diagonal of the measured imprint, confirms the amplitude of the transformations observed by the metallographic observation and the X-ray diffraction.

We conclude that through a simple methodology of hardness measurement, it is possible to precisely evaluate the growth kinetics of the carbide layer. This method can be validly transposed to diffusion - precipitation studies.

**References**

[1] Pancielejko M, Pecht W and Czyzniewski A 1999 Tribological properties of PVD titanium carbides *Vacuum* **53** pp 57-60
[2] Sundgren J E, Jacobson B E and Hibbs M K 1984 Microstructure of nitride and carbide coatings prepared by physical vapor deposition methods *Z. Met. kd.* **75** pp 855-861
[3] Bromark M, Larsson M, Hedenqvist P, Olsson M, Hogmark S and Bergmann E 1994 PVD coatings for tool application: Tribological evaluation *Surf. Eng.* **3** p 205.
[4] Shikama T, Shinno H, Fukutomi M, Fujitsuka M, Kitajima M and Okada M 1983 Properties of $Ti_xC_{1-x}$ films coated on molybdenum by magnetron sputtering *Thin Solid Films* **101**(3) pp 233-242
[5] Wolfe D E and Singh J 2000 Titanium carbide coatings deposited by reactive ion beam-assisted, electron beam-physical vapor deposition *Surf. Coat. Technol.* **124** pp 142-153
[6] Storms E K 1967 *The Refractory Carbide* (New York: Academic Press)
[7] Toth L E 1971 *Transition Metal Carbides and Nitrides* (New York: Academic Press)
[8] Varacalle Jr D J Lundberg L B Herman H and Bancke G 1996 Titanium carbide coatings fabricated by the vacuum plasma spraying process *Surf. Coat. Technol.* **86** p 70
[9] Wendler B and Benarioua Y 1993 Vacuum annealing study of thin Ti layers on high carbon steel substrates *Pro. Int. Conf. on Applied Crystallography (Katowice)* World Scientific (Publishing Co. Pte. Ltd) pp 228-230
[10] Benarioua Y, Lesage J, Bemporad E and Chicot D 2006 Titanium carbide films obtained by conversion of sputtered titanium on high carbon steel *Surf. Coat. Technol.* **200** pp 5447-5454
[11] Benarioua Y, Boubaaya R, Lesage J and Chicot D 2013 Annealing study of thin chromium layers on cemented steel substrates *Surf. Coat. Technol* **227** pp 65-69
[12] Meyer E 1908 *Phys. Z.* **9** p 66
[13] Chicot D and Lesage J 1995 Absolute hardness of films and coatings; *Thin Solid Films* **254** pp 123-130

**Institute of Physics**
Dirac House, Temple Back
Bristol BS1 6BE UK

ISSN: 1742-6588
ISBN 978-1-5108-6408-5

CURRAN ASSOCIATES INC.
proceedings.com